How Attention Works

How Attention Works

Finding Your Way in a World Full of Distraction

Stefan Van der Stigchel
Translated by Danny Guinan

The MIT Press
Cambridge, Massachusetts
London, England

This publication has been made possible with financial support from the Dutch Foundation for Literature.

N **ederlands**
letterenfonds
dutch foundation
for literature

This book was set in Stone Sans and Stone Serif by Jen Jackowitz. Printed and bound in the United States of America.

Library of Congress Cataloging-in-Publication Data is available.

ISBN: 978-0-262-03926-0

10 9 8 7 6 5 4 3 2 1

Every one knows what attention is. It is the taking possession by the mind, in clear and vivid form, of one out of what seem several simultaneously possible objects or trains of thought. Focalization, concentration, of consciousness are of its essence. It implies withdrawal from some things in order to deal effectively with others, and is a condition which has a real opposite in the confused, dazed, scatterbrained state.

—William James

Contents

Preface

In 1995, a 27-year-old police officer, Kenny Conley, found himself pursuing a suspect after a shooting incident in Boston. Other police officers were also involved in the chase. When these officers spotted a man climbing over a fence, they immediately apprehended him. They used considerable force when doing so, and the man suffered kidney damage and serious head injuries. However, there was an even bigger problem: the man they had arrested was not a criminal but an undercover agent who had also been pursuing the suspect. In the ensuing investigation into the possible use of excessive force, Kenny Conley was called as a witness. He had arrived on the scene while the suspect was being arrested and must have seen which officers were involved. However, Conley insisted that he had not seen the struggle and had been focused purely on apprehending the fleeing suspect. The jury didn't believe him. They were convinced that he had attempted to protect his colleagues when giving his statement. Conley was charged with perjury and obstruction of justice and sentenced to 34 months in jail. He appealed the decision and was released from custody pending the outcome of the appeal. After seven years of legal wrangling, he was finally acquitted and awarded damages of $600,000.

Researchers decided to take on the challenge of recreating the events that led to Conley's initial conviction. Test subjects were asked to pursue a suspect along a sidewalk and to count the number of times the suspect touched his head with his hands during the chase. After reporting this information to the researchers, the test subjects were then asked whether they had noticed the fight that had been staged eight meters away from the sidewalk while they were chasing the suspect. Most of the test subjects had failed to see the fight, just like Kenny Conley.

If you dig deeper into the subject of visual perception, you will quickly discover that we actually register very little of the visual world around us. We think that we see a detailed and stable world, but this is just an illusion created by the way in which our brains process visual information. This has important consequences for how we present information to others—especially attention architects.

Everyone whose job involves guiding people's attention, like website designers, teachers, traffic engineers, and, of course, advertising agents, could be given the title of "attention architect." Such individuals know that simply presenting a visual message is never enough. Attention architects need to be able to guide our attention to get the message across. This means that they have to fight for our attention: website designers use all kinds of tricks to capture the attention of visitors, magicians fool us by distracting us, and directors manipulate our attention to make their movies appear as realistic as possible. Whoever can influence our attention has the power to allow information to reach us or, conversely, to ensure that we do not receive that information at all.

After reading this book, you will hopefully be more aware of the continuous battle for our attention that rages all around us. Attention architects try every trick in the book: they hang up screens with moving images in busy city centers, add flashing banners to websites, and develop computer programs with blinking icons to attract our attention. Sometimes attention architects do this for our benefit and sometimes for their own, like when they want to sell us something. We have a certain level of control over our attention, and so we can try to ignore their attempts to distract us. But we don't always succeed. We are often slaves to our own attention system and can become distracted at times when we really should be focusing our attention elsewhere. This is why it is important to know exactly how attention works.

1 Barrier? What Barrier? The Illusion of a Rich Visual World

There was genuine cause for celebration in the Netherlands on July 21, 2014, when the Coen Tunnel, one of the country's worst traffic black spots, was reopened after major renovation work. It was hoped that the revamped tunnel would help traffic in the area to move more smoothly. One of the improvements was the addition of an extra tunnel that can be opened and closed, depending on the volume of traffic. When not in use, this tunnel is closed by means of a clearly visible barrier. To prevent motorists from accidentally driving into it, they are warned of the tunnel's closed status by a series of bright red crosses above the road that can be seen from a good distance out. If you were to drive up to the Coen Tunnel yourself, you would find it hard to believe that anyone could fail to see the barrier.

Less than a year after the tunnel was reopened, a 63-year-old motorcyclist suffered serious injuries when he failed to notice that the extra tunnel was closed and crashed straight into the barrier. It was the twentieth such accident since the renovation job. The authorities tried to warn road users more effectively by hanging up additional warning signs, using vehicle-mounted flashing arrows, and installing steel traffic cones. In addition, the barrier itself was fitted with flashing LED lights and made to appear larger. The warnings were all to no avail. Drivers continued to crash into the barrier at high speed without even taking their foot off the gas.

Of course, one explanation is that some road users are distracted by their radio or mobile phone at the critical moment when they should be heeding the warning signs, but this does not apply to every accident. After all, the barrier can be seen from a long way out and drivers tend to watch the road carefully. And in the case of the unfortunate motorcyclist, it is highly unlikely that he was busy checking his phone or tuning the radio. So why do things still go wrong?

When we go for a walk in the woods, we generally enjoy the sight of all the trees around us and the myriad shades of green. We open our eyes and allow our visual environment to work its magic. Our eyes are our window to the world. All we have to do is open them to be able to see. It happens automatically. Spotting a squirrel in a tree or following the tracks of a horse are reflex actions. We believe that what we see is the whole picture: stable, rich, and vastly superior to any virtual environment.

However, we actually take less information on board from our surroundings than we might think. For example, movies are full of continuity errors that viewers usually fail to spot. Very few of us ever take any notice when a jacket that was hanging on a coatrack is suddenly not there anymore in the next scene. The legendary *Star Wars* movies are famous for these kinds of mistakes. Objects move from one position to another, and a background full of plants and trees suddenly changes into a barren desert. You only ever notice these discrepancies when someone takes the trouble to point them out to you, with the result that it is almost impossible not to see them the next time. Of course, movie directors do their best to keep such mistakes to a minimum, but the fact that neither they nor the people who edit their movies manage to spot these errors in the first place demonstrates just how easy it is to miss them.

The infamous "gorilla clip" is now so well-known that I can't even use it in my psychology classes with first-year students anymore. But in case your memory needs refreshing, here's a quick recap: Two groups of students are throwing a basketball back and forth, and the viewer's task is to count the number of times the team with the white T-shirts throws the ball. At a certain point, a gorilla walks into the frame. He beats his chest with his fists and then walks out of shot again. The majority of those seeing the clip for the first time fail to spot the gorilla. When I recently decided do this exercise again with my students, they believed that I would be showing them the all-too-familiar clip. So they paid special attention to the gorilla with the aim of showing their lazy professor that he should know better than to try to fool them with the same old trick again.

However, what the students did not know was that I had shown them a new version of the clip, one in which the curtains hanging behind the basketball-playing students gradually change color and one of the players walks abruptly out of the frame. The effect is even greater than in the original clip. Almost none of my students noticed either of these two major changes, primarily because they were too busy waiting for the gorilla to appear. Fooled again.

The fact that we cannot register everything we see in our visual world leads to frequent claims that the human visual system is an inefficient and flawed one. After all, it seems pretty inefficient not to be able to spot a gorilla when it appears on-screen.

I disagree with this hypothesis, however. We may occasionally fail to notice major changes in our visual world, but just how big of a problem is that really? Our brain works on the assumption that the world around us is a stable and consistent one, which is usually the case. Curtains don't change color and objects don't move suddenly from one place to another. And even if they did, it would not really be a problem if you failed to notice. It's all about gathering the information that is relevant to you. That is what you need to focus on. You can ignore all of the stuff that is of no value to you. A system that tries to process every scrap of visual information it receives is cumbersome and inefficient. There is simply no need to process all of the information available to us. Okay, you didn't spot the gorilla, but your job was to count the number of passes, not to look for gorillas. And you succeeded in doing just that.

Systems that use less energy have an advantage in the evolutionary scheme of things. An efficient system makes the energy it does not use available to other systems, and that is what our visual system does, too. The retina catches the light from everything around us, but only that information which is relevant to us is processed. The image of the gorilla falls on your retina, but you ignore this information because it is not relevant to your needs. And that's a good thing. Imagine yourself walking around in a supermarket trying to process all the information you see. You would know the brand and price of every single product and that would cost you far too much energy.

Close your eyes for a moment and picture the place in which you are reading this book. How much detailed information can you provide about the space around you? Maybe the space is familiar to you, and so you can recall certain details from memory, but if you are in less familiar surroundings, then the level of detail you will be able to provide will be pretty low. You could then justifiably conclude that we have a very limited internal representation of our external world. The visual system possesses a unique feature that allows it to represent information very selectively:

our continuous access to the visual world. All of the visual information that is available to us at any given moment is 100 percent accessible. All we have to do is to open our eyes, and the information floods in. This means that we can use the visual world as a kind of external hard drive. We don't need to store every single detail related to the external world in our internal world because all of the visual information is continuously available to us externally.

The only thing that we need to be able to recall internally in order to interact effectively with the external visual world is where the relevant information is located in relation to our own location at any given moment. For example, if you want to find out what color T-shirt the person sitting next to you is wearing, all you need to know is where this person is sitting in relation to you. You then turn your eyes to that spot and immediately see the color of the person's T-shirt. There is no need for you to store every detail of the visual scene you see before you in your internal memory.

Imagine the following: you and a friend are walking down a busy street in town on your way to a coffee bar at the end of the street. There are people everywhere and neon signs flashing all around. At that moment, only certain aspects of the visual world are relevant to you: the coffee shop in the distance and your friend walking beside you. You are moving, so all of the information is moving too, relative to your position. You use your eyes to access the visual world around you and note only the location of the information that is relevant to you. A gorilla could walk down the street, or everyone could pull on a different T-shirt, but you would not notice it. However, you will notice if the coffee bar suddenly disappears, your friend runs off, or the gorilla starts screaming because that information is relevant to you. You can afford to ignore everything else.

This feature of our visual system is also responsible for the illusion that we enjoy a rich visual world. You don't know what you are missing, and you can't miss what you don't know. If someone were to tell you that while you were walking down the street with your friend all of the clothing shops had suddenly turned into shoe shops, you would not believe them (just like the students who initially didn't believe that I had managed to fool them again with the gorilla clip). However, in most cases, changes like these or the things we miss are not pointed out to us. It is only when we walk into a lamppost or crash into a traffic barrier that we are confronted with how little we actually take on board from the world around us.

An even more important factor in the maintenance of the illusion of a rich visual world is the way in which we monitor it. Think of the light in a fridge: it is always on when we open the fridge and we can never be 100 percent sure that the light goes out when we close the fridge. You can only check this by opening the fridge again, and when you do, the light is always on. It works much the same way in the visual world, too. If you want to check the richness of your perception, you can focus on a particular object and experience it in full. But your surroundings could change completely while you are doing so, without your even noticing! And when your focus shifts to the next spot or object, you once again have a specific and rich experience, and so on. When we are out walking in the woods, we are aware of the richness of our perception, but we can never fully experience every single tree that we see at any given moment.

In the example of the coffee bar, you probably thought there was no way you would not notice the gorilla walking calmly down the street. We might have an efficient visual system, but in an evolutionary sense it would also be very useful if we had the capacity to spot the gorilla. A gorilla in the middle of the street can spell the kind of danger that suddenly makes your trip to a coffee bar seem far less important. The same applies when you see a car speeding toward you. As we will see in the following chapters, there are all kinds of situations in which you automatically pick up visual information. This information may not be relevant to the task with which you are currently preoccupied, but it may be very relevant to your ability to avoid danger. Fortunately, the visual system is designed to allow this kind of information to "interrupt" our focus on task-specific information. However, you will still probably fail to spot the gorilla walking down the street if it doesn't start screaming or waving its arms. Just like in the infamous clip.

There are, of course, exceptions to the rule that we are often unable to recall the details of a visual scene at a particular moment. Take Stephen Wiltshire, for example. He is able to draw an extremely detailed panorama of a cityscape even when he has been flown just once in a helicopter over the city. Stephen—also known as "The Human Camera"—is an idiot savant, a person with outstanding cognitive abilities in one particular field. These people generally either are autistic or suffer from a mental retardation but are still capable of performing one specific task extraordinarily well. Stephen, for example, did not start talking until he was nine years old, yet he was able to draw extremely intricate sketches of buildings at the age of seven. The level of detail he can recall is mind-blowing, but this seems to happen at the expense of other skills, such as language. We don't yet know exactly what goes on in the brains of people like Stephen. Temple Grandin, a renowned scientist who herself suffers from autism, believes that many autistic people think in images instead of words. This may partly explain the extraordinary visual skills of Stephen Wiltshire.

Regardless of how selective our perception may be, we are still bombarded with visual information every day. We see screens everywhere informing us about train timetables, special offers, or the latest sports results. Our telephones have screens; our computers have screens. When it comes to the transfer of information, the visual system is our single most important sensory tool. When road construction work is being carried out and the public needs to be informed, this is generally done using visual information. If you wish to convey information about road closures through auditory channels (although this is very unlikely given how well soundproofed cars are these days), you need to use speech. It takes a lot longer to convey the same information orally through speech than visually with the aid of symbols. This is because the visual system is able to process information in the blink of an eye. If you show someone a very detailed photo for just a second or two, they will still be able to describe the image to you fairly accurately afterward.

This ability to rapidly process information is clearly demonstrated in the experiments conducted by Mary Potter in the 1970s. She was interested in the speed at which people are able to process the information present in a single scene. Test subjects were given a written description of a scene (e.g., "traffic on a street") and then asked to find that scene among a series of images presented to them in quick succession. They were instructed to press a button as soon as they had identified the scene that matched the written description. No visual information regarding the scene was provided—neither the color of the cars nor the layout of the street. When the test subjects were shown eight scenes per second, there was a success rate of 60 percent when it came to finding the scene that had been described in writing to them. Eight scenes per second means that each image was visible for just 125 milliseconds, and the test subjects had to process all of the visual information in each scene within this extremely short space of time. In the case of a second group of participants—who only had to describe which scenes they had seen after the event—it is probably not surprising that only 11 percent of them were able to describe the scenes in any detail. Although they could say which scenes they had been shown, they were unable to provide any specific information about the content.

While it may be impossible to fully register everything we see in our external visual world, Mary Potter's experiments demonstrate that we are able to take in an entire scene literally in the blink of an eye. This apparent contradiction can be explained by establishing what we mean by "seeing." All of the visual information that falls on the retina is registered by our brain, too. This information includes the colors and shapes that make up the world around us, and it is processed in the primary visual cortex. However, at this stage we are still not able to identify the individual objects. "Seeing" describes everything that falls as light on the retina. While we might "see" a lot of stuff, we only process a small amount of the information deeply enough to know what that stuff actually is. Identification—that is, knowing whether something is a tree or a green building—requires more in-depth processing and access to the identity of the object. I will go into this in more detail in chapter 3, but for now it is sufficient to know that the illusion of a rich visual world has less to do with shape and color and more with our knowledge of what a particular object is precisely.

If we define "seeing" as the information that is picked up by the retina, this means that we "see" automatically and that simply "seeing" something does not require any further processing.

Back to the traffic barrier at the Coen Tunnel. There can be no doubt that while the unfortunate motorcyclist was approaching the tunnel, the visual information about the barrier fell on his retina. This means that its color and shape must have been processed by the primary visual cortex. So the motorcyclist did "see" it. However, he failed to identify the barrier as a barrier. To do that, he would have to have processed the information about the barrier at a deeper level, and this did not happen. The same applies to the gorilla: everyone who watches the clip "sees" the gorilla, but because viewers are preoccupied with counting the number of times the ball is thrown back and forth, they do not process the visual information about the gorilla deeply enough to identify the gorilla as a gorilla.

This also explains our ability to recognize scenes that are shown to us very briefly. To do so, we use the basic visual information that we are able to pick up almost instantaneously. This information provides us with the "gist" (the basic contents) of a scene. However, the information is not available long enough to allow us to identify individual elements in a scene. As a result, we cannot describe the scene in any great detail, but we can identify a scene that was described to us in advance.

If you want to communicate a visual message, such as the information on a traffic sign, it is important to know what kind of information you can communicate in an instant. When I am cycling in town, I am often amazed by the number of yellow signs crammed with complicated instructions about ongoing road construction work. Visual information can be communicated very quickly, but there are limits. We are not able to process full sentences in the blink of an eye. Symbols are much more effective in this regard. Of course, I realize that it is impossible to devise a symbol for every piece of information, but when a road has multiple complicated signs, it cannot be but to the detriment of both the message being communicated and the intended recipients of the information, that is, the road users.

The communication of information is said to be successful when the relevant information reaches the intended user. Regardless of how impressed we might be by a particular advertisement, if we do not remember the intended message after seeing the advert (e.g., the name of the product), then the advertisement will not have worked and the attention architect will have failed in his or her task. The creator of an advertisement actually has two tasks: to ensure that it is seen and that the relevant information is communicated. This is especially important in today's world in which every strategy imaginable is employed by those trying to communicate their information to us. Jet fighter pilots have access to a lot of information about their aircraft, but it is crucial that their attention is captured immediately when a problem like engine failure occurs. In the case of a road crowded with signs, one of the signs might be trying to communicate very important information, however ineffectively. It is up to the attention architect to make sure that the information is communicated properly and that important information is not only seen but also processed at the identification level.

When communicating information, attention architects also have to consider the fact that we are all different. People are now living longer, and many elderly people find it difficult to keep up with the current speed at which information is transferred. Smartphones might be perfect for the current younger generation, but that generation too will eventually grow old (and probably a lot older) and develop a need for a less hurried manner of information transfer. The trend now is to encourage the elderly to live independently in their own homes for longer, and so we will have to adjust our manner and means of communicating information to suit their needs as well—for example, in the design of smartphones.

We are using more devices with screens than ever before now, and manufacturers continue to try to make them more and more user-friendly. However, for some firms the user-friendliness of a device is not the only consideration. Using the device must also be "an experience" in itself. Not only must it be easy to use, but you should also feel better as a result of using it. This means fancy colors and modern design techniques. But user-friendliness and slick design do not always go hand in hand. The mark of a really successful designer is the ability to combine both of these aspects and to come up with devices that are suitable for all age groups.

Quick question: have you read this book up to now without putting it down, or did you stop at any stage to check your e-mail? Many of us are addicted to checking screens that are constantly trying to provide us with information. We check our telephones, tablets, and computers frequently for new info. More often than not, I discover that there is no new information, but that doesn't stop me from checking again five minutes later. I even become a little nervous when I turn off my e-mail because I need to concentrate on the task at hand. Sometimes I prefer to be interrupted by a visual alert telling me that I've got mail, just in case something pops up in my mailbox without my noticing it. Thankfully, my symptoms are still relatively mild, but the popular media has already coined a term for this new type of addiction: "infobesity." In fact, it is increasingly being referred to as a clinical disease. The term is the brainchild of a "trend team" employed by a company specializing in identifying trends among young people. Although there is some doubt as to whether or not it can be classified as a disease (there is very little scientific literature on the subject), the fact is that doctors are treating more and more teenagers these days for problems associated with a lack of sleep.

One of the factors contributing to this lack of sleep is our insatiable appetite for information that is presented to us on-screen. Unsurprisingly, this leads to problems with concentration. I would probably have finished this book a lot sooner if I had shut down all the irksome sources of information around me, but I just wasn't able to do so. This is one of the reasons why many people prefer to work at night, a time when others decide for us that there will be less new information doing the rounds, making it easier for us to concentrate on what we are doing without fear of interruption. Somehow or other we are often incapable of turning the information tap off ourselves, which is usually the result of our "fear of missing out," that is, the fear of missing new information, especially from people within our social network. It is not difficult to see why young people have more trouble with this phenomenon than the elderly, simply because the social media network is far more important to the younger generation.

From the scientific studies that have been carried out, we know that young people are extremely frequent multimedia users. On average, 18-year-olds spend a total of 20 hours a day on various media. Of course, this can only be because they use different media simultaneously. One particularly interesting finding is that the majority of this multimedia use is of the visual kind: clearly, we have a preference for all things visual. Functions that rely on the spoken word have been replaced by visual ones: the voicemail is fast becoming a thing of the past because it takes too much time, and people prefer to send their messages screen-to-screen instead. We are using the telephone less and less and choosing more often to interact with others on-screen and not only through hearing their voice. If e-mail and WhatsApp relied on the spoken word, they would be a lot less popular.

Screens are so efficient at communicating information that we see them everywhere nowadays. The result is a titanic battle for our attention. We have already established that it only takes a quick glance at a limited amount of visual information to know what that information is. In a single moment, we choose the one piece of visual information that is most relevant to us from all of the information swirling around us. We then process this one piece of information deeply enough to be able to establish its identity. All of the other information continues to blink away furiously, but it can only become relevant when we decide to look again.

This knowledge of the limited processing capacity of visual information has consequences for artificial systems like robots. There is a lot of debate about the role that robots will eventually play in human society. Although they are nowhere near us yet in terms of intelligence, the expectation is that robots will soon be taking over many of our simpler tasks, for example, robot vacuum cleaners that can clean our living room or office without our intervention. In the more distant future, robots will be able to travel around entirely on their own—Google's self-driving car is a good example of how close this distant future may actually be. The Google car is already driving around in Nevada, albeit with two humans on board: one who can take control of the car if something goes wrong and one who monitors the car's performance on a laptop. In June 2015 it was revealed that there had been 23 accidents involving these autonomous cars, but that in all cases it was human and not robot error that had been the cause. The accidents had been the fault of either the driver of another car on the road or the human driver of the Google car.

If you were designing a robot, it would be a good idea to draw your inspiration from the very efficient human visual system. At any given moment, we process only a very limited amount of visible information to a level that enables identification. Our choice is determined primarily by what is most relevant to us at that moment. If you are driving a car, the things that are directly related to driving on a public road are the ones that are of most relevance to you. You can ignore everything else. During this process we can suddenly be presented with information warning us of imminent danger: a child running across the road, for example. How many times have you slammed on the brakes because out of the corner of your eye you spotted a child about to run across the road? Even though you were focused on the road, at that moment you reacted automatically because your brain detected something that spelled possible danger. More about this later on.

An efficient robot that doesn't waste time on unnecessary calculations would probably behave in a similar way by processing only the visual information that is relevant to its task while at the same time being aware of any incoming information that may be of importance. All of the other information that is picked up by the robot's camera can be ignored and does not need to be processed. Although a robot that can perform an unlimited range of tasks equally well may still be far off in the future, the road is definitely (and literally) being paved for robots that can focus on one specific task while navigating in traffic. And of even more benefit, perhaps, is the fact that the Google Car is not distracted by incoming e-mails or WhatsApp messages and is therefore potentially far safer than the eternally curious human. As long as our Google Car can successfully spot a traffic barrier looming in the distance, we will be able to sit back and check our e-mail to our heart's content while the robot carries on doing what it was designed to do.

Our perception is the result of an efficient system that selects certain visual information for further processing. This principle applies not only to the real world but also to the virtual world that we find ourselves in when we put on a virtual reality headset. If we only process that information which is relevant to us at any given moment at a deep level, then it is probably unnecessary to provide people with very detailed information about everything they see when they are in a virtual environment. This significantly reduces the amount of processing power that is required, one of the major stumbling blocks today in the development of virtual reality environments.

Virtual reality has always been something of a holy grail, but significant developments have been made over the past few years and expectations are now very high. Many large technology companies, such as Facebook, Sony, and Valve, have begun to manufacture relatively inexpensive and lightweight headsets that allow users to experience virtual environments. Facebook's Oculus Rift is a good example of a headset that has met with widespread acclaim. It is relatively inexpensive, weighs less than 450 grams (approximately one pound), and offers a 100-degree field of view. While the headset still only offers a fraction of the resolution and field of view of the human eye, it is a massive improvement on previous models. Older models were not only much heavier but also tended to make users feel nauseous, a problem that has long hampered the development of virtual reality systems. These systems' low temporal resolution (number of frames per second) made users feel sick, but the faster graphic video cards that are available today now offer a solution to this problem. And there are even ways of turning a smartphone into a virtual reality headset, such as Google Cardboard.

The possibilities offered by virtual reality are almost endless. It is already being used to relieve pain in surgical procedures, such as the treatment of third-degree burns. Placing the patient in a virtual environment of snow and ice can have a pain-relief effect similar to that achieved by morphine. Virtual reality can also allow traumatized soldiers to revisit combat situation experiences without any risk. And the uses also extend to education: fancy attending a lecture on the theory of relativity given by Einstein himself?

At the moment, the makers of these systems are endeavoring to create as complete a virtual image as possible by maximizing the level of visual detail in a single scene. However, this demands a very high level of processing power from the computer controlling the headset to produce a degree of visual detail that we don't actually need. In addition, we are only able to focus sharply on something with a very small part of the retina, that is, the fovea. Look straight ahead of you and try to read something on the periphery of your vision. You won't be able to. A smart system only needs to provide information with a high degree of visual detail about the spot upon which the eyes of the user are actually trained. As we will see later on, this particular spot is usually the place where the information that is relevant to the viewer is to be found.

This is exactly what so-called gaze-contingent multiresolutional displays do. These are virtual reality headsets that present information with a high visual resolution only at the spot where the user is looking. All other information is displayed in low resolution. The system does this by monitoring the viewer's eye movements during use of the virtual reality headset. It is rumored that this function will soon be a standard feature in all future Oculus Rift headsets. If developers then focus their efforts solely on what users look at and adjust the image they build accordingly, it will be possible to solve many of the remaining problems with virtual reality systems. It will no longer be necessary to display excess information in sharp detail. The resulting savings on energy could be as high as 80 percent. This would also lead to a longer battery life, which would be very useful when we want to remain in our virtual world while being driven around in an automatic car, for example.

So, even though we do not pick up everything in our visual world, there are many ways in which we can predict which information we can expect to receive in any given situation. However, there are limits to the ways in which attention architects can convey information. And installing a traffic barrier on a highway is never a good idea, regardless of how eye-catching one tries to make it. It's all about expectation: we simply don't expect to encounter a barrier on a highway, and that makes it extremely difficult to warn drivers adequately. To a large extent, what we see is determined by what we expect to see. Regardless of how salient an object may be, if it doesn't match our expectations, we will more than likely fail to see it. It would probably be better to spend more time and energy on building a better tunnel than on trying to get road users to notice your barrier.

2   Why Fire Trucks Shouldn't Be Painted Red: What Makes Something Stand Out?

The Ladbroke Grove metro station is located in West London and is part of the Great Western Main Line, an important railway connection that runs from Paddington Station to the west of England and southern Wales. On October 5, 1999, a train departed from Paddington Station at around 8 a.m. Just before it reached Ladbroke Grove station, the train was supposed to stop for a red light so that it wouldn't end up on the wrong track. Unfortunately, the driver didn't stop. Instead he carried on at full speed and crashed headlong into an oncoming train. The diesel tank of the oncoming train exploded and resulted in a fireball that destroyed several carriages on both trains. The casualties numbered 31 dead and 520 wounded.

The above tragedy may appear to be very similar to the story about the Coen Tunnel traffic barrier in the previous chapter. Both cases involve a failure to notice a visual signal. And yet they are very different from each other. In the case of the Coen Tunnel, the barrier is an unexpected element and the primary cause of the problem is the fact that road users do not expect to encounter a traffic barrier in the middle of a highway. However, the same does not apply to a red light. After all, one of a train driver's most important tasks is the correct interpretation of signals.

Unfortunately, the 31-year-old train driver did not survive the accident, meaning that the cause of the tragedy can never be fully determined. However, it is considered likely that the accident was caused by the erroneous interpretation of a signal and the driver's resulting belief that he could carry on safely down the line. But how could he make such a mistake? The red signal for the track he was traveling on was visible directly above the rails on the left of a line of four other signals for the adjacent tracks. All of these signals were red at the time, meaning that the accident could not have been caused by the train driver's reading the wrong signal.

There must have been another reason for the driver's fatal interpretation. One possible cause is the visibility of the signal in question. On that day in October, the weather had been particularly fine, with the sun shining brightly in the sky behind the driver. Although the light emitted by the signal would have been no different than on any other day, the bright sunlight may have made it difficult for the driver to determine whether the signal was actually red or not. Most of us are familiar with the difficulty of establishing whether our car's headlights are working on a very bright day. The contrast with everything else around us is reduced considerably when there is a bright sun shining in the sky, compared to when it is dark.

What makes the situation even more complex is the actual construction of the signals themselves. They contain both red and yellow lights, and when the red lights are shining, the yellow ones are turned off. However, the low October sun directly behind the train may have resulted in a strong reflection from the yellow lights, making it difficult to establish whether it was the red or the yellow lights that were illuminated. In this case, the reflection from the yellow lights was probably almost as intense as the light from the red signal. To add to the problem, the visibility of the signals was hampered by the electricity transformers hanging above the rails.

Incidentally, this was not the first time that a train had driven through a red light at this particular spot. In the six years previous to the fatal accident, drivers had erred on no fewer than eight separate occasions when they were supposed to stop at signal SN109. The railway company knew about these incidents, but they did not take the appropriate steps to correct the situation. Proper training and instruction on the most hazardous signals may have been enough to prevent this accident from occurring. The unfortunate driver had only completed his training 13 days previously, but this had not included any instruction on signals that were prone to being ignored.

The fact that a red signal can be clearly seen one day but completely missed the next has nothing to do with the light source itself but rather with the place where that light source is located. An object that is "visible" in one situation is not necessarily so in a different situation. The term "visible" deserves some further explanation. We use our eyes to inspect the details of objects around us. When we move our eyes, we focus the middle of the retina—the fovea—on the object we wish to inspect. The fovea is the part of the eye that is most sensitive to detail. However, this does not mean that everything upon which we train our eyes is actually visible. The small letters at the end of a contract are often barely legible, even when you try hard to read them. A detail has to be big enough in order to be visible. For example, at a certain distance it becomes difficult to make out the individual letters in this book.

The contrast with the background also plays an important role in the visibility of an object. This is readily demonstrated in the legibility or otherwise of the text in figure 2.1.

This manner of text presentation might look stylish, but the contrast between the letters and the background is much lower than if you were to use black letters on a white background. The greater the contrast between an object like a letter or a red signal and its background, the greater the visibility of the object.

When a popular TV listings magazine in the Netherlands decided to change its design in 2008, it was inundated with complaints from subscribers who said they couldn't read the program information anymore. The font size had been made smaller, and the information (such as the channel and recording codes) was now being printed in various tints of gray. Some of the text was so light in color that the contrast with the white background was minimal. Elderly people in particular were not happy with the new design, and it wasn't long before the makers figured it would be best to reverse the changes.

An object that is visible is not necessarily salient. A salient object must be not only visible but also instantly distinguishable from its surroundings. For example, a letter in a book will stand out enough (i.e., from the page) to be clearly visible, but it will not be salient because there will be hundreds of other letters on the same page, too. However, a single letter printed in the middle of a white page in a book is very salient.

The army uses camouflage so that its soldiers won't stand out against a green background. These days, however, many wars are fought in urban environments or out in the desert, and so the standard military uniform no

It is currently fashionable for websites to use dark gray text on a light gray background.

Figure 2.1
The degree of contrast between lettering and background affects the visibility of lettering.

longer uses green but rather gray-colored camouflage. In a combat situation in the desert, a soldier wearing green camouflage is not only visible, but also sticks out like a sore thumb because of the difference between the green uniform and the colors of the desert. The ideal soldier's uniform would be one that automatically changes color to suit its environment, like a chameleon. Whether an object can be termed visible or salient depends on a number of other factors, such as how it contrasts with the background and the amount of light illuminating the object. In the case of the fatal train accident in Ladbroke Grove, it is quite possible that the reflection of sunlight on the yellow lights in the signal made it difficult for the driver to see whether the signal was actually red or not and that this is what led to the resulting tragedy.

If you want an object to be identified more quickly, you have to make the object more visible *and* more salient. From a young age, children are taught that fire trucks are red. Indeed, when people are asked to name a typically red object, chances are that the first thing they will say is "a fire truck." But is the color red really the best choice for a fire truck? Even when you take into consideration the dangers automatically inherent in driving a fire truck to the scene of an emergency, the number of accidents involving these vehicles is still rather high. When fire trucks first appeared on our roads, there weren't that many red cars driving around, but today the picture is very different. Red is now a very common color among cars, and that means that fire trucks don't stand out as much as they used to anymore. Of course, there are other ways of alerting road users to the presence of a fire truck, such as by using sirens and flashing lights. Another clever and popular device is the addition of yellow reflectors or white/blue stripes.

A more radical option would be to paint fire trucks a completely different color altogether. Such a change would have to be accompanied by a very effective information campaign so that people could get used to the fire truck's new look as quickly as possible. In the United States, a number of states have already taken this step, and their fire trucks now have a lemon-yellow color, which is not a color seen very often on or near a public road. In 1997, the fire department in Dallas, Texas, put both red and yellow fire trucks out on the road so that they could monitor the number of traffic accidents in which each type was involved. And the result? The lemon-colored fire trucks were involved in far fewer accidents than their red counterparts. The bright yellow color was much more conspicuous, and other road users were quicker to react to the presence of a lemon-colored fire truck.

In the Netherlands, ambulances also have a yellow color (RAL 1016) that is very similar to the color used on many new fire trucks in the United States. It is, after all, very important that ambulances stand out from the crowd. Other road users are prohibited from driving around in a vehicle with a color that closely resembles that used by the emergency services' ambulances. Not so long ago, animal ambulances in the Netherlands were forced, under the threat of a hefty fine, to change their color because they were too similar in appearance to normal ambulances. And in 2012 a Dutch court ordered a security company to remove the stripes from its company vehicles because they bore too much of a resemblance to those used on the nation's police cars. Despite the fact that the combination of colors was different from that used by the police, the court ruled that the thickness and the direction of the stripes, and the white background on which they were painted, could result in confusion among other road users as to the vehicles' true identity. Interestingly, during the case a police spokesman suggested to the security company that they paint sunflowers on their vehicles instead. He was probably only trying to be helpful, but sunflowers were unlikely to do the security firm's tough public image much good.

There are plenty of other examples of how increasing the salience of important information can result in fewer traffic accidents. One example is the so-called third stop lamp: a brake lamp that is located at eye level on a car's rear window.

In the Netherlands, this stop lamp is now mandatory for cars built after the year 2000. The third lamp is much more visible because it is located at a point higher than the other two regular brake lamps. Motorists can also see this third lamp not only on the car immediately in front of them but on other cars as well. In the United States, it is estimated that the third stop lamp has reduced the number of traffic accidents each year by almost 200,000.

The limits on our perception are also an important consideration in the construction of tunnels. During the day, motorists find themselves entering a darkened space when they drive into a tunnel. There is a huge difference in terms of intensity between sunlight and the lights in a tunnel. On a clear, sunny day our visual system is adjusted to the bright light of the sun, and it takes us a moment or two to adapt to the darkness of the tunnel. Those few moments can be crucial. A similar problem presents itself when we reach the other side of the tunnel. Of course, the lighting in a tunnel compensates for this, with the lights at the entrance and exit usually being brighter than those in the middle. This makes the transition less abrupt and gives the visual system more time to adjust to the changed circumstances.

Let's return to the red fire truck for a moment. The use of the color red can be a problem for people who are color-blind. Color blindness is more common among males; it is estimated that one out of every 12 males experiences difficulty with identifying colors correctly, as compared to one out of every 250 females. Humans detect colors by using three types of photoreceptor cells in the retina, known as cones. Color blindness is usually caused when one or more of these three types of cones cease to function properly. The most common form is red-green color blindness, whereby a person has difficulty distinguishing between red and green tints of color. This makes it very difficult for someone to distinguish between a red fire truck and a gardening firm's green truck based solely on color.

Although color blindness is not categorized as a disease, it can be extremely inconvenient for those who suffer from it. Everyone is familiar with the scene in blockbuster movies in which the hero has to disarm a ticking bomb. He receives an instruction in his earpiece to cut the red wire and not to touch the green one. We can only hope that our hero does not suffer from red-green color blindness. It also plays a role in the more mundane, everyday activities of an electrician, for example. The old color coding for electrical wiring used green and red for the phase and neutral wires. Nowadays, the colors brown and blue are used for these wires in order to eliminate any problems that might arise because of color blindness. For the same reason, another very important wire, the earth or ground wire, has been given two colors so that even those people who cannot identify any colors at all will still be able to see the difference. It is prohibited to give other wires the same kind of two-color identification.

Traffic lights also use the colors red and green. However, people with red-green color blindness generally have no problem knowing when to stop and when to go. This is because traffic lights make use of a phenomenon called double coding: the status of the traffic light is indicated not only by the color of the light but also by its location. Red is always above and green is always below. In Belgium they have devised a system, one that supposedly eliminates all doubt, in which the color of the red traffic light has a more violet hue and the green traffic light has a slightly blue tint. You might be asking yourself now if the accident at Ladbroke Grove was possibly caused by color blindness? The answer is "No" because all train drivers are tested for color blindness, just like pilots, driving instructors, and soccer referees.

Double coding is not the only useful tool when it comes to helping people who are color-blind. Back in the days when TV broadcasts were in black-and-white, the shirts that soccer teams wore had to differ in tint from each other. Usually, one was dark in color and the other light so that the viewer could tell who was who when watching the game. It is also a useful device for color-blind soccer players, as the different tints of teams' shirts helps them to see which players they should and which players they should not pass the ball to.

How different people experience different colors can vary greatly. In February 2015, a photo of a dress appeared on the Internet, and within 24 hours it had gone viral. If you haven't seen it yourself yet, simply Google "the dress" and you'll find it. The photo was taken by a proud mother who wanted to show her daughter the dress she was planning to wear to her upcoming wedding. However, the couple-in-waiting couldn't agree on the color of the dress. She saw a white and gold dress, while he insisted that the dress was blue and black. So the daughter decided to enlist the help of her friends and put the photo up on Facebook. The rest is history: literally millions of people got involved in the discussion, and there were soon two very clear-cut camps: the white-and-gold camp versus the black-and-blue camp.

One of the most interesting things about this story is that it didn't seem to matter how the people involved viewed the photo; differences of opinion arose even between people who scrutinized the dress on the same mobile phone or laptop. And it appears that they could only see one color combination or the other, unlike what happens with a Necker cube whereby a person's interpretation of what he or she sees can change from one moment to the next. In the case of the dress, a person sees one color combination from the very start. All around the world, color specialists were woken from their beds at night by journalists hoping to solve the puzzle. A fierce debate ensued between scientists everywhere as to the explanation for this phenomenon. It soon became obvious that the key to solving the puzzle lay in the concept of color constancy.

Color constancy allows us to distinguish the same colors under various conditions of illumination. For example, a yellow banana looks very different when viewed in a room with blue lighting compared to when it is lying outside in the sun. When we look at the banana under the blue light, our visual system takes the color of the light source into consideration and "peels" it (excuse the pun) from the actual color of the banana. Our visual system draws on our knowledge of the color of bananas to ensure that we continue to see the banana as yellow regardless of how it is illuminated. In the case of the dress, it is clear from the photo that the light is coming from an outdoor light source, but it is not clear whether that source is "golden" sunlight or a "blue" sky. It is this ambiguity that results in the different interpretations. When a person's visual system assumes that the light is coming from a blue sky, the blue layer will be peeled from the person's perception and the dress will appear to be white/gold in color. On the other hand, when a person's visual system assumes that it is sunlight that is illuminating the dress, the gold layer is peeled off, leading to the conclusion that the dress is blue/black in color. This entire process happens at the unconscious level and is extremely difficult to influence.

A good example of how we compensate for variations in illumination is Edward Adelson's optical illusion (figure 2.2) using a chessboard with light- and dark-colored squares. The square containing the letter *A* appears to be darker than the one containing the letter *B*. However, they are both exactly the same shade of gray. Our visual system's job here is to determine the color of each square. In doing so it takes into account the shadow thrown by the large object in the illustration. As a result, we end up with a completely different interpretation of the shade of gray of the square upon which the shadow falls. Just look at the image on the right in which the

Figure 2.2
Optical illusion.

relevant section is shown in 2-D. When the sense of depth disappears, so too does our interpretation of the shadow, which means that the visual system no longer has to compensate for its effect. We can clearly see that the two squares are exactly the same shade of gray. Note that this is not a fault on the part of our visual system. On the contrary, it shows just how efficiently we deal with the various circumstances in which we encounter objects in our world.

The case of "the dress" demonstrates that there can be huge differences between our perceptions of the same object and that those perceptions can be heavily influenced by our previous experiences or the opinions of others. A simple disagreement regarding a light source can result in two people's perceiving an object completely differently. There are many different research groups currently busy studying this particular phenomenon. At a recent scientific congress on the subject of perception, the exact same dress was the most worn item of clothing among the female researchers in attendance. All the hullaballoo regarding the dress certainly didn't do its sales figures any harm, that's for sure. And by the way, the dress is really blue and black.

The salience of an object depends on a variety of factors, in particular the difference between the object and its surroundings. This can be a matter of color or where the object is located in our field of vision. As we have already discussed, we cannot focus sharply out of the corner of our eye. We use the retina, which is located at the back of the eye, to catch the light of objects around us. The retina contains photosensitive cells called rods and cones. The cones allow us to see color and to focus on objects, and most of them are located in and around the fovea, a spot 1.5 millimeters (approximately 0.06 inches) in diameter where our focus is at its sharpest.

When we observe the world out of the corner of our eye, we find it more difficult to distinguish between individual objects because our focus is not as sharp there. Three simple lines drawn closely together are easy to distinguish when you look at them directly (i.e., when the foveae of both eyes are focused on the lines), but it's a lot more difficult when you look at them out of the corner of your eye. The difficulty of distinguishing between objects that are aligned closely together is known as "crowding," and there are a few simple rules that govern its severity. The first rule is probably the most logical one: the more a set of objects is moved to the corner of your eye, the more difficult it becomes to distinguish the objects from one another. Look at the plus sign in the illustration below, and, at the same time, try to identify the middle letter of the three letters on the right. The greater the distance between the plus sign and the letters, the more difficult the task becomes. This is because the further away from the fovea an object is presented, the greater the drop in focus. You have to make an object appear much larger in the corner of your eye if you want to keep the same level of saliency.

+   x Z x

+   x Z x

The camouflage example has already shown us that an object will stand out less when it is surrounded by similar-looking objects. This is the second rule of crowding. Look at the illustration below and try to distinguish the middle letter in each line. You will see that this is more difficult on the second line where the Z is flanked by capitals. The three letters in the second line are all the same size and appear more similar to each other than the letters in the first line where the x's are shown as lowercase letters.

+   x Z x

+   X Z X

Designers of typefaces are very conscious of the fact that letters are more difficult to read when they appear too similar to each other. Special typefaces have been developed by designers who claim they are easier to read for people with dyslexia. Letters can be made to look less alike by making the internal spacing in the letters a little bigger and/or by slanting them slightly. People who suffer from dyslexia often confuse their b's and d's because the only difference between them is that they are mirror images of each other. Altering the b and the d by giving the letters slightly different contours could, in theory, make it easier for people to distinguish one from the other. While these ideas are certainly interesting, the jury is still out, in scientific circles at least, on whether they can actually help people with dyslexia to read more easily or whether they are nothing more than another commercial sales pitch.

Letters on traffic signs are often made from reflective material, and when these letters are illuminated at night by the headlamps of passing cars, the individual lines shine so brightly that the letters can start to look a lot like each other. An e and an o are both round, and the only real difference between them is down to two little lines. In the United States most of the traffic signs seen on the highway now use a typeface called Clearview. This typeface purports to solve the problems that result from brightly shining letters by using thinner lines and more subtle variations between the letters. Place names are also no longer displayed in capital letters, but instead with the first letter of the word in uppercase and the rest in lowercase. Studies have shown that the Clearview typeface has enabled people to recognize words at distances up to 16 percent further away than was possible with the previous typeface. At an average driving speed of 90 kilometers per hour (about 56 miles per hour), this represents a gain of two seconds in terms of recognition time.

For all typefaces, an important rule is that individual letters should never be too close to each other. This is also the third rule of crowding. The closer objects are to each other, the more difficult it is to distinguish them out of the corner of your eye. Look at the two lines below and see for yourself.

+   xZx
+   x Z x

If you are an attention architect and you are trying to present information to people in a way that will catch their eye, you have to play by the rules of crowding. Logos are usually located in the corner of advertisements on billboards so that as much space as possible can be kept free for the message. However, when the logo is placed too close to other visual information, its visibility may be reduced significantly. The logo also needs to be different in appearance from the rest of the information on the billboard. These days, most TV channels display their logo in the corner of the TV screen, and they are usually easily distinguishable against the background. Sometimes they are not static but instead spin continuously in the corner of the screen. It is not surprising, therefore, that many viewers find these logos incredibly annoying.

People with advanced macular degeneration (MD) are not able to use the fovea to view information closely. Macular degeneration causes the cones in the fovea to die off, thereby severely impairing the patient's vision. The fovea, which, as mentioned above, is located in the center of the retina, is essential to the ability to perform tasks such as facial recognition and reading that require sharp focus. In the advanced stages of MD, degeneration leads to the forming of a blind spot (or scotoma) in the visual field while all of the other areas outside the fovea remain more or less intact. The field of vision of an MD patient is like a field with a blind spot right in the middle, precisely at the location where a healthy eye can focus most sharply.

Macular degeneration affects people of all ages. Juvenile macular degeneration is often the result of a genetic disorder caused by Stargardt disease. This form of degeneration affects about one out of every 10,000 people in the population. For people over the age of 50, the figure is much higher: about one out of every 50.

People who have this kind of blind spot compensate for the loss of vision at the center of the visual field by using the corners of their eyes to perform the tasks normally reserved for the fovea. They pick a spot on the retina that serves as their "new fovea," also known as a pseudofovea. You may have encountered this before in people who, instead of looking at you directly, seem to do so out of the corner of their eye. Their fovea, the part of the eye you would normally use to look straight at someone, is looking at something else entirely. This is logical, of course, because the person looking at you can see next to nothing with that part of their eye. Various studies have shown that patients who switch to using a pseudofovea can read much better than those who do not select a new permanent location for the fovea. As a result, much of the therapy currently being applied to alleviate this condition focuses on helping patients to find a spot on the retina that can best serve as a pseudofovea. It is useful to know, for example, that a location in which the blind spot appears on the right-hand side of the visual field is less suitable for people who wish to improve their reading ability. This is because most people (in the Western world) read from left to right and often use the visual information on the right to speed up the process of reading.

A person who uses a pseudofovea experiences more difficulty with crowding than a person with normal visual acuity because of their inability to use the fovea to inspect the relevant information closely. Unfortunately, our brain is not so flexible that it can enable the pseudofovea to take on the characteristics of the real fovea or facilitate a training effect through regular use of the pseudofovea, thereby alleviating the problem of crowding. This means that MD patients always focus using a spot in the eye that does not provide optimal vision. Reading, in particular, becomes very difficult because of the problem of crowding: the letters appear very close to each other and even look alike, and for MD patients this spells double trouble, as they have to try and read these letters literally out of the corner of their eye.

Attempts have been made to find ways of presenting text so that it causes less crowding, thus making it easier to read with a pseudofovea, but they have met with little or no success.

There are two more problems when it comes to using a pseudofovea. First of all, we generally don't like people looking at us out of the corner of their eye. Someone who doesn't look us straight in the eye can come across as uninterested and ambivalent, especially if we are meeting that person for the first time. And even if we have known someone who suffers from MD for a long time, it still feels like an unnatural way of communicating when the person appears to be avoiding our gaze. We tend to form a negative opinion of someone we think is unwilling to look us straight in the eye. Secondly, our bodies are built in such a way that our eyes are naturally meant to focus straight ahead. Looking at the world through a pseudofovea can lead to neck pain as a result of having to constantly turn one's head.

In an attempt to solve these problems, scientists have developed prism glasses with lenses that are polished so that they can project the image that would normally fall on the fovea onto the corner of the eye. A person wearing a pair of these glasses can continue to look straight ahead while still making use of the pseudofovea. The glasses do not improve the wearer's eyesight, but they do help to counter the negative effects of looking at the world through a pseudofovea. Unfortunately, the glasses are still very heavy, which makes it difficult to wear them for long periods of time.

While wearing a pair of prism glasses certainly requires some initial adjustment on the part of the wearer, this is usually not a problem in the long term. Those of us who wear reading glasses already know what it is like when we start wearing a new pair. Everything looks a little strange at first: for example, the lenses might be more concave or convex so that the visual information falls slightly differently on our eye; the frames might be different and appear to obstruct our vision at first. However, after a few days we will have become so used to our new glasses that we will even forget we are wearing them.

These kinds of adjustments can also take on very extreme forms. In 1950, Theodor Erismann conducted a bizarre experiment in which he instructed his assistant to wear a pair of glasses that turned the whole world upside down. At first the poor assistant could barely function normally. He couldn't walk down the stairs without falling, couldn't pick up objects that were lying right under his nose, and was barely able to walk properly.

However, after a few days the assistant started to adjust to the new situation—so much so, in fact, that after ten days he was so used to his upside-down world that he could carry out all his everyday tasks without any trouble at all. He was even able to ride a bike. Our brains are good at adapting to new situations, even to the extent that the new situation can quickly become the "normal" one.

We are obviously able, therefore, to adapt to changes in our visual perception. This also applies to our diminishing eyesight as we get older. Not so long ago in the Netherlands there was a vociferous lobby for the introduction of mandatory driving tests for people above the age of 45 who wished to renew their driving license. A government minister even submitted proposals for the necessary legislation in 2004. The thinking behind the proposed changes was that older people posed a danger to road safety because of their diminishing eyesight. It was assumed that they were often unaware of the problem and could potentially cause more accidents as a result. Other countries in Europe began to discuss the topic, too. Eye doctors were particularly vocal in their support for the proposed changes. Of course, new legislation would prove very lucrative for this branch if it were to lead to mandatory eye tests. In the Netherlands alone it would mean an extra half a million tests per year...

You would think that good eyesight is absolutely crucial to safe driving practices, but the scientific literature on the subject tells a different story. It turns out that there is almost no correlation between a person's eyesight and the number of traffic accidents the person causes. In fact, some studies suggest that people with less than average eyesight are actually safer drivers. This is because of the adjustments we make based on our own ability to see stuff around us. When people cannot focus very sharply, they will probably drive a lot more carefully. Think of a foggy evening, for example, when all road users, not just those with bad eyesight, adjust their driving behavior to suit the difficult conditions. A mandatory eye test would therefore make no contribution to safer driving. Fortunately, the minister heeded the advice of the scientific community and decided to withdraw the proposals.

In the first chapter I arrived at the conclusion that we are only able to represent a limited portion of the external world in our internal world. This second chapter has shown us that the system we use to realize that representation also has its limits. For example, we are only able to focus sharply with our fovea, and we also experience problems with crowding under various circumstances. In addition, external factors such as illumination can have a major effect on how we perceive color. And because we blink our eyes regularly, they are often closed and therefore unable to register all of the available visual information.

We usually don't notice the effect of these limitations on our visual system because of how flexible and fast that system is. Take the blind spot on the retina, for example. The retina converts the light that falls on the eye into electrical and chemical signals that are then sent to the brain for further processing. These signals travel from the eye via a spot where a large network of axons (nerve fibers) connects to the optic nerve. There are no cones or rods at this spot, which is why we are blind to the visual information that falls on this part of the retina. This blind spot is about four visual degrees in size—about the width of four fingers at the end of an outstretched arm. You can find your own blind spot using this book and the illustration below by closing your left eye and looking at the plus sign with your right eye. Extend your arm and bring the book slowly toward your eyes. At a certain point—approximately 20 to 25 centimeters (about 8 to 10 inches) from your face—the circle will disappear. At that moment the circle is situated in your blind spot, and the visual system automatically fills up the empty space with the color of the page (a process known as "filling in"). The circle then reappears as you continue to draw the book closer to your eyes. This filling in happens even when you are looking at a complex visual stimulus like static on a TV screen.

+   O

Phenomena like crowding mean that an attention architect needs to remember that not all intended visual messages are detected by the viewer. An object has to be salient enough to be able to stand out from its surroundings. But how do you make something salient? What makes it stand out? Color would appear to be the most obvious starting point, but why is that?

Our visual system is designed to process a number of basic visual properties separately from each other. For the purposes of clarity, from this point on I will refer to these visual properties as "building blocks" because they are the basic ingredients of the visual system. In order to fully comprehend the building-block concept, it is important to establish what is meant by a visual field. When you look straight ahead, you are looking at the middle of your visual field. All information on the right-hand side is situated in the right visual field and the information on the left-hand side is situated in the left visual field. The information that is collected by both eyes is sent along the optic chiasm (a point of intersection in the brain) to the visual cortex, the part of our brain that processes visual information. During this transport, the images from both eyes are combined: the visual information from the left visual field of both eyes is sent to the right half of the brain while the information from the right visual field travels to the left half. The visual cortex contains neurons that only react (or "fire") when information is presented in a certain part of the visual field (the receptive field). These areas also contain neurons that only react when a certain type of visual information is presented in the receptive field. There are neurons that are specialized in movement and orientation—for example, is the line slanted to the right (/) or to the left (\)? David Hubel and Torsten Wiesel won the Nobel Prize in Medicine in 1981 for discovering that certain neurons in the visual cortex only fire when a line with a particular angle (e.g., /) is presented in the receptive field. All of these building blocks are then put together in the visual system to form whole objects that we are able to see.

The information on all of the various building blocks is passed on to areas in the brain that are specialized in processing these blocks further. For example, color is processed in an area called V4. When individuals suffer damage to this color area, they are no longer able to perceive color and can only see their world in shades of gray, a bit like watching a black-and-white TV. This color area can also be identified in healthy test subjects by exposing the area to a short magnetic pulse using a technique called transcranial magnetic stimulation, which briefly disrupts the color processing mechanism.

The perceptibility of an object is determined by the extent to which the building blocks, such as color, shape, and size, are distinguishable from their surroundings. The greater the difference, the more likely we are to notice the object. Perceptibility often plays a decisive role in the eternal battle for our attention. Up to now I have avoided using the word "attention" unless strictly necessary because I first wanted to explain how our brains process the basic visual building blocks. But now the question has to be asked: what is attention exactly? In the next chapter we will see that it is attention that ultimately determines what we gather from the visual world around us.

3 Selection through Attention: Why We Like to Stare at a Blank Wall When We're Thinking

In April 2006, a woman decides to pay a visit to her doctor in Amsterdam. She has discovered a lump in one of her breasts and is a bit worried. The doctor does an x-ray, and the results reveal a tumor. Two months later the woman has a lumpectomy to remove the tumor and undergoes a period of radiation therapy. The odd thing about the situation is that six weeks before she went to see her doctor she had already been x-rayed as a part of a nationwide screening for breast cancer among women aged 50 to 74.

Although she feels perfectly fine in the wake of her surgery, one question continues to trouble her: why did nothing show up during the screening? The woman submits a complaint to the relevant authorities in Amsterdam. They tell her that the radiologists did not make a mistake and that her complaint is being dismissed. The radiologists had labeled the anomaly detected during the screening as a "minimal sign," meaning that they believed there was no need for any additional diagnosis.

Studies in the Netherlands have shown that this screening procedure has a detection probability of around 70 percent. This means that radiologists fail to detect incidences of cancer in over one quarter of all women who do in fact have breast cancer. This score is not ideal, of course, but neither is a situation whereby every single possible cancer diagnosis, no matter how minimal, has to be followed up by an additional examination. These examinations are very painful, and reacting to every minimal sign would lead to a lot of unnecessary discomfort and stress. The chances of detecting a tumor on the basis of a minimal sign are known to be very low.

Generally speaking, public trust in medical screenings is shaky to say the least: when mistakes are made, the willingness of the public to take part diminishes, with the result that the government fails to achieve its primary objectives. The verdict of the authorities in Amsterdam was therefore very important to the continued existence of medical screenings. Some radiologists in England and the United States are now refusing to take part in screenings because of a fear of being sued for malpractice. In the United States, legal firms are starting to handle more and more claims for damages on behalf of patients. They use TV advertising to alert patients to the possibility of suing hospitals for negligence. In the Netherlands, on the other hand, the number of such cases is relatively low. This difference between the United States and the Netherlands is reflected in the number of women who are referred to a specialist after the first scan: ten out of every hundred women in the United States compared to two out of every hundred in the Netherlands.

When the scans in which a tumor was missed are checked again, the tumor usually turns out to be visible. In those cases, the radiologist who initially viewed the scan either failed to spot the tumor or interpreted the results incorrectly. This may appear shocking at first, but that is not necessarily so. Compare the two situations again. The radiologist who does not know whether a scan contains a tumor will interpret that scan as he or she would any other. The low statistical probability of finding a tumor means that the radiologist's expectation of spotting one will be similarly low. However, when the scan is assessed for a second time, the situation is completely different. The radiologist now knows that the scan does in fact contain a tumor and there is a maximum probability of actually finding it. As a result, the tumor will almost always be detected the second time around. It would therefore be wrong to regard the second radiologist as better at his or her job than the first one, as their entirely different circumstances make it impossible to compare one with the other. The first radiologist would more than likely spot the tumor if he or she were to check the scan again, simply because the probability of that happening would be much higher.

When assessing a scan, a radiologist has to carefully study all of the relevant areas in all of the x-rays. To do this properly, a radiologist must inspect each of the scans individually, as well as all of the individual areas within each scan. Spotting a tumor is not like picking Santa Claus and his red suit out of a crowd of little helpers all dressed in green. Neither the color nor the shape of a tumor is all that much different from the color or shape of the surrounding tissue. In fact, it is extremely difficult to detect abnormal tissue, and to be able to do so requires many years of training and extensive medical knowledge. You might reasonably expect radiologists, therefore, to be absolute experts at inspecting the visual world. They wouldn't miss a gorilla if it walked into the shot, would they?

In an attempt to answer this question, a study was carried out in which a group of radiologists was asked to inspect a number of lung scans for traces of malignant tissue. What the radiologists did not know was that an image of an angry gorilla had been added to the scans. The image was about the size of a matchbox. An angry gorilla is not the kind of visual element you would normally expect to find in a scan, like the situation with the gorilla in the basketball clip. The results of the study were astonishing: 83 percent of the radiologists failed to spot the gorilla in the scans.

The outcome can be explained on the basis of the actual task that the radiologists were asked to carry out. They were looking for tissue with specific visual characteristics, not for a gorilla. If the color and shape of the image of the gorilla had resembled in any way the color and shape of malignant tissue, it would probably have been spotted by more of the radiologists. This kind of information is crucial to the training of professionals whose job demands the ability to search meticulously for certain visual objects. It is also important to the work of security personnel who have to be able to scan camera images for possible safety threats. A person who is instructed to search for a specific object will usually do so to the exclusion of all other objects. There has even been a case reported in which a group of emergency room radiologists, interns, and doctors were unable to find a guide wire left behind in a patient's vein despite the fact that it was clearly visible in three different computed tomography scans. It appears that what you are likely to find depends mostly upon what you are looking for (this applies to lots of different things in life, but here I am referring only to the visual act of searching).

Of course, it's not only *what* you are searching for that is important but also *how* you search for it. The same scientists who carried out the study with the gorilla and the x-rays also looked at how radiologists go about inspecting scans. By tracking the eye movements of radiologists, the scientists were able to reveal that they could be split into two different categories: "drillers" and "scanners." Drillers select one specific location on the screen, which they then search on all of the different scans. When they are finished searching that section, they move on to a new one and check all of the scans again. Scanners, on the other hand, search each scan completely and meticulously before moving onto the next scan. They examine each scan once and only once. Radiologists tend to use one or other of these strategies exclusively in their work.

The study showed that the drillers searched a larger area of the scans than the scanners did and also that they scored better when it came to locating abnormal tissue. In the future this kind of information will be very valuable in the training of radiologists. It is becoming increasingly easier and less expensive to measure eye movements, which will make it possible to train radiologists to be more efficient in their search methods, and also to intervene when a radiologist is thought to be searching scans incorrectly or incompletely. Radiologists' eye movements can reveal which parts, if any, they have overlooked in a scan.

Scans are also used at airports for checking hand luggage, and security scanner operators spend hours every day searching the contents of bags and suitcases. For security reasons, little is known about the techniques they use to check the scans, but there are reports that show that operators fail to spot up to 75 percent of the fake explosives that are sometimes hidden in bags for test purposes. Of course, the chances of finding explosives in a scan are much smaller than the chances of a radiologist's spotting an area of abnormal tissue. The intermittent insertion of fake explosives may help to raise the probability a little, but it still does not prevent operators from regularly failing to spot the real thing.

An American study into the performance of airport security personnel revealed that having to scrutinize scans on a daily basis helps them to be more precise when carrying out other unrelated search work. Out of a group of test subjects who were asked to find a well-hidden object on a computer screen, 82 percent were successful. A group of professional security scanner operators, however, scored 88 percent for the same test although they did take longer to complete the task compared to the nonprofessionals. This study showed that the professional security personnel were more precise in their searching methods, which can probably be attributed to their many years of experience monitoring luggage scans. However, they cannot be called "superexperts" because their greater precision required more search time. Compared to the nonprofessional test subjects, they searched longer and continued searching when others would be more inclined to give up.

One of the problems with these laboratory experiments is that the object the subjects are required to find is usually present, which is not always the case in the daily work of security scanner operators. Given that the chances of finding an object depend in part on the probability of the object's actually being present, the results of these experiments do not tell us all that much about the operators' working methods. This is clearly evident in the case of security scanner operators who were tested during their last week of training. They failed to spot the hidden objects during periods of testing in which relatively few objects were hidden in the scans compared to periods in which the objects were present more often. This is why some airports have already introduced the practice of inserting images of forbidden items into operators' screens on a regular basis. This increases the probability of a hidden object being found, meaning that the operators are more likely, in theory at least, to spot forbidden items hidden in luggage.

If you would like to check your prowess as a security scanner operator, you can download Airport Scanner, a free app that sets players the task of finding dangerous items in luggage scans. It has been a huge success worldwide and has millions of users. The app is partly funded by the American government, which is naturally delighted with the amazing amount of information they are able to glean from the game. Researchers are also involved in the development of the game, and the first scientific articles were recently published containing the data retrieved from one billion searches. Some players have become so addicted that they have already completed thousands of searches. This, in turn, has provided developers with the opportunity to insert certain objects only at very sporadic intervals (in less than 0.15 percent of the searches). This is exactly the kind of thing you cannot test in a laboratory because your test subjects would eventually end up running screaming from the lab after being subjected to hours and hours of tests. Based on a probability of 0.1 percent, an object will appear once every 1,000 searches, and in order to reach any firm conclusions about a player's performance when attempting to find an extremely rare object, you need to carry out at least 20,000 searches. This kind of data is now available thanks to the Airport Scanner app, and it has proven beyond doubt that players/professionals frequently fail to spot these rare, hidden objects.

When an x-ray is being scanned, the image is usually examined centimeter by centimeter for abnormalities. But what happens, exactly, during this process? To answer this question, I must now properly introduce the term "attention." Attention is the mechanism we use to make a selection from all of the visual information available to us and then to process only that information. All of the information upon which our attention is not focused is by and large filtered out. Attention can be compared to the neck of a bottle when the contents are being poured out: only a certain amount of liquid can pass through at any given moment.

There is a lot of discussion surrounding the term "attention" in the field of psychology. This is primarily because of the lack of a standard definition. Everyone knows that the concept of "memory" is related to the ability to retain information. And a concept like "perception" obviously has something to do with how we receive sensory information. With attention, however, it is not so clear-cut. If I tell people I meet at a party that I am currently researching the area of attention, I usually hear major differences in what they believe attention actually means. Initially, most people say that it relates to our ability to persevere with a particular task, and they say things like "My attention keeps drifting" or "I have a short attention span." This interpretation of attention describes the ability to choose a certain activity, like reading this book, and then to persevere with that activity. To be able to do this, you must not allow yourself to be distracted by other matters or perform other activities, like thinking about what you need to get at the grocery store later on. This interpretation is all about selection, but then including a time component in which it is important to persevere with an activity for a certain period of time. This has little to do with the kind of attention I am talking about in this book, that is, our attention for visual information. Nevertheless, in both cases we are talking about attention.

This lack of a proper definition is a problem not just at parties but also within the field of study itself. Many researchers are of the opinion that it is impossible to provide a standard definition of attention because the concept is used to describe so many different selection processes. Some scientists even believe it would be better to do away with the term altogether as it seems to give rise to more questions than it does answers. One person's interpretation of the term "attention" can be entirely at odds with the next person's. The result of this ambiguity is often seen in scientific debates where instead of discussing how to interpret the outcome of a certain experiment, scientists get bogged down in an interminable discussion about the term "attention."

The absence of a good definition of attention can have far-reaching consequences. People who have suffered brain damage often experience difficulty carrying out normal, everyday activities. When a patient continuously forgets information, this is treated as a problem associated with memory loss. Patients with brain damage are also often diagnosed as having an attentional deficit. An attentional deficit can point to a patient's difficulty with concentrating on a single task, but it can also refer to problems with visual attention. However, in the absence of a standard definition of attention, the question is whether such a diagnosis is more of a hindrance than a help to the patient.

Thanks to the emergence of new techniques for studying and mapping the brain, such as fMRI (functional magnetic resonance imaging), we are beginning to find out more and more about the functions of the various parts of that mysterious organ. It is interesting to note that many of these parts are now regarded as very important to the functioning of attention. However, given the hazy definition of the term, no one has yet been able to identify exactly what the unique functions of these areas are. The solution appears to lie in discontinuing the idea of attention as one single concept and regarding it instead as a general term that describes the various selection mechanisms used by the brain. Our brain requires the use of selection simply because we are unable to perform multiple tasks, process all of the visual information around us, and think every single thought we could possibly think all at the same time. We select pieces of this information and process them without allowing ourselves to be distracted by irrelevant information. Selection ensures that the brain does not become overloaded by having to process all of the information available to us simultaneously.

When we talk about the selection of visual information, we are actually talking about visual attention, which concerns the processing of incoming visual information only and has nothing to do with one's ability to persevere with one single task. It has to do with the information that you process at any given moment. This is still a pretty vague definition, however, and it doesn't really help us to clarify anything. We still don't know, for example, what it means when you focus your attention on a particular spot in space. In order to explain this properly, we need to carry out a thought experiment.

Imagine a world in which there are only two objects: a red square and a red circle. We know already from the previous chapter that different building blocks are processed by different parts of the brain. When processing the information from this imaginary world, the color of the objects—red— is processed in the part of the brain that is responsible for processing color, while their shapes—round and square—are processed in the part that is responsible for processing shape. The "color neurons" react to the red color and the "shape neurons" fire in response to the square and the circle. The visual system has no trouble combining this information: there is only one color after all, meaning that both the square and the circle must be red. Reassembling the building blocks in this abstract but simple world is not a problem.

We will now make this simple world a little more complex by adding a second color so that we have a red square and a blue circle. In the color part of the brain, the red and blue neurons become active, while the situation in the shape area remains as it was. But which color goes with which shape? When this information has to be combined, the visual system has no way of knowing which shape is red and which is blue. The system knows that it has seen a red object and a blue object, but it does not know whether the square is red or blue. This is referred to as a "binding problem" and is a consequence of how the visual system functions. The information about building blocks that we receive from our rich visual world cannot be combined because these building blocks are processed in different neural areas (to say nothing of the fact that our visual world contains far more shapes and colors than I have used in the simple, abstract example above).

Visual attention offers a solution to this binding problem. Attention ensures that only a certain part of the visual world is processed; the rest of the information is filtered out. In the above example, our brain solves the binding problem by focusing its attention on the red square and ignoring the blue circle. This enables the visual system to know which pieces of visual information should be bound to each other. The neurons that react to "red" fire in the color area, and the neurons that react to "square" fire in the shape area. So the visual system now knows that there is a red square at the spot where the attention is being focused.

In order to know what a particular object actually is, you need to have access to the information on all of the object's combined building blocks. Only attention can ensure that the information on an object's various building blocks is brought together and bound. A green tree will be nothing more than the color green and a random collection of shapes unless attention is focused on the spot where the tree is located. Attention binds the color and the shapes to each other and allows us to experience the tree as one single object. This means that we have no access to the identity of objects if our attention is not focused on them.

The binding of building blocks is a continuous process. It happens extremely quickly and automatically and cannot be interrupted. When we focus our attention on something, the building blocks at that spot are immediately bound to each other. This is also the reason why we like to stare at a blank wall (or close our eyes) when we are thinking hard about something. It offers us temporary respite from extraneous information that could otherwise disrupt our thought process.

Henk Barendregt, a winner of the Spinoza award (the most prestigious award for scientific research in the Netherlands), has carried out extensive research into mindfulness. When I told him about the automatic characteristics of the binding of information, he commented that one of the most important tools in mindfulness meditation (*vipassana*) is the dissolution of the links between one's experiences and one's thoughts. This kind of meditation represents an effort to influence the binding process and to experience the world in an unconditional manner by ceasing to automatically bind pieces of information. The objective is to be able to experience the world, including your own thoughts, in an uninhibited manner, with the ultimate goal of diminishing or even eliminating all suffering that arises from attachment.

In the following chapters I will examine in more detail the cases of people who experience problems with visual attention as a result of damage to their parietal cortex, but in the meantime it may be of interest to point out that these patients tend to report significantly more incorrectly bound objects than do people without brain damage. They have less attention for a certain part of the visual field. This absence of attention means that they cannot bind the various building blocks properly to each other, which results in an increased number of incorrectly bound objects.

It is also possible to create incorrectly bound objects in healthy test subjects. In a series of experiments, visual information was displayed on a screen for a very short period of time (e.g., 200 milliseconds) and then masked by a screen with visual noise. This masking prevented afterimages from appearing on the screen at the moment when the visual information was withdrawn. By doing this, the researchers could be sure that the information was visible for exactly 200 milliseconds, no more and no less. The short duration ensured that the test subjects did not have enough time to move their attention from one spot on the screen to another. In that short burst they saw a screen with two numbers and four objects. The four objects consisted of a combination of various building blocks, such as a red triangle, a small green circle, a large yellow circle and a large blue triangle. The test subjects were first asked to say which numbers they had seen and then which objects. In this situation, where the visual information had been displayed very briefly and attention was also required to identify the numbers, the test subjects did not have enough time to focus their attention on the individual locations of the four objects. The result was that the test subjects reported incorrectly bound objects. In 18 percent of all cases they reported an erroneous combination of the size, color, and shape of two different objects and gave answers like "a small red circle" or "a small green triangle." In situations where the numbers were not displayed, they did not report any incorrectly bound objects. Without the numbers, therefore, they had enough time to move their attention from one object to another and to bind the various building blocks to each other. The results are convincing evidence that attention is indeed responsible for binding building blocks to each other.

During theater performances, spotlights are often used to light up certain parts of the stage. This helps the audience to focus on the spot where the action is taking place at critical moments. At these moments the actors in the spotlight are more visible than everything else on the stage. This situation also gives stagehands the chance to change the decor on parts of the stage where the spotlight is not shining without the audience noticing. A spotlight is also an excellent metaphor for visual attention: attention enables the deeper processing of visual information, and visual information that is not the focus of attention is largely ignored.

As with many metaphors, the spotlight is an oversimplification of how attention actually works. For example, a spotlight travels across a stage, but the same cannot be said of attention. Attention can be focused on one point for a moment, only to shift to a completely different point a moment later; this happens without any attention being paid to the locations between these two points, unlike a spotlight that generally stays lit when moving from one spot to the next. However, it is possible to keep your attention trained on an object as it moves through space: for example, a ball rolling across the floor.

This particular metaphor is very suitable in terms of variations in the size of the spotlight. A spotlight can be made larger or smaller, as can the attentional spotlight. The size of that spotlight—the so-called attentional window—is an important feature of visual attention and one that we can control. If we have to perform an activity for which we need to know the identity of a small element, like a letter, we can make the spotlight small. Look at the illustration below. When you focus your eyes on the plus sign and try to read the letters at the same time, you will have to make the spotlight smaller and shift your attention from one spot to the next. This is the only way to gain access to the identity of the individual elements. If, however, you only need to know where the different letters are located, you will not need to use a smaller spotlight and can therefore make your spotlight larger. In that case you will not have access to the identity of the individual letters, but you will still be able to pinpoint their location.

$$
\begin{array}{ccc}
 & H & \\
G & & J \\
F & + & K
\end{array}
$$

The extent to which detail is required to be able to perform a task determines the size of the spotlight. The illustration below shows what is known as a Navon figure. This is a large letter made up of smaller letters. If I showed you a Navon figure very briefly and then asked you to identify the large letter, you would have no problem doing so. This is because in preparation for the task you would have already increased the size of your spotlight. However, if, immediately afterward, I asked you to identify the small letter used to make the large letter, you would probably not be able to tell me. To see that kind of detail, you would need to use a much smaller spotlight. Although you saw the small letters and your eyes registered the visual information, the identity of the small letters will have escaped you because you were registering only the identity of the object upon which your attention was focused.

HHHHHHHHH
H
H
HHHHHHHHH
H
H
HHHHHHHHH

Most of us know an elderly driver who has been the cause of a dangerous situation because he or she failed to pay attention to the information in the corners of his or her eyes. Or an older pedestrian who takes no notice of anything or anyone around him or her anymore and seems to have a much smaller visual world. These problems can be partly attributed to the fact that our spotlight becomes smaller as we get older. The capacity of the spotlight is called the "useful field of view" (UFoV). The fact that our UFoV grows smaller with age means that it becomes more difficult for us to register the entire visual field around us. Various studies have shown a relationship between the size of the UFoV and the number of traffic accidents or the time it takes to cross the road at a busy intersection. There are training programs whose aim is to increase the size of the UFoV. In these programs, participants are asked to identify objects displayed to them at various points on a computer screen for steadily decreasing periods of time. Results show that this kind of training can be very beneficial: it leads to fewer traffic accidents and keeps elderly people active longer.

Elderly people are not the only ones with a smaller UFoV. Young children also experience difficulty in taking in the entire visual world around them. Recently, a simple observation made this very clear to me. My five-year-old son and I decided to have a day out together at a mega indoor playground where there was a terrific ninja training game. The game involved hitting a button as quickly as possible in reaction to the lights that appeared on the periphery of your vision. I was surprised at how slowly my son reacted to the flashing lights. It was like he studied the lights one by one to see which one was on, while I was able to tell immediately. My reaction time was five times faster than his, and I'm no ninja. I later found out that my conclusions were supported by scientific experiments that show that children don't have as expansive an overview as adults do. It also turns out that they react a lot quicker to the smaller letters in Navon figures than they do to the large letter. This can be very useful information, especially when you have difficulty in understanding why your child cannot find his or her favorite teddy bear even though it is lying right there in front of them on the ground: children (almost literally) don't see the bigger picture.

When radiologists are searching for abnormal tissue, they will make their spotlight smaller and focus their attention only on those areas that are important, which also makes it easier for them to miss a gorilla hidden in a scan, of course. This phenomenon is known as "inattentional blindness": blindness caused by a lack of attention for a particular spot while the hidden object is plainly visible.

Head-up displays (or HUDs for short) are becoming increasingly popular nowadays—for example, in the airline industry. A similar device, the HMD (head-mounted display), uses a pair of glasses or a helmet that projects information in the field of view of a driver or pilot. Displays like these eliminate the need to look away from the road or sky. One disadvantage of these displays is that you can be so busy processing the projected information that you miss other important incoming information. Experiments have shown that even experienced pilots with thousands of flying hours sometimes fail to spot another aircraft when using a HUD in a simulator, even when that aircraft is right in front of them on the runway! The danger with these devices is that one's attention can become divided between the runway (or the road) and the projected information, even when that information is not important at that particular moment.

But does that mean that we just have to accept the fact that we will always fail to spot certain things going on around us? No, not necessarily. There are many factors that can help to alleviate inattentional blindness. For example, people who play a lot of basketball are better at spotting the gorilla in the infamous clip because they do not need to focus so much of their attention on counting the number of times the ball is passed. And test subjects also spot the gorilla more often when the color of the gorilla is the same as the colors worn by the team they have been instructed to monitor. In this case, the level of inattentional blindness is reduced because the object that the test subject has been asked to monitor has visual characteristics that match the task being carried out by the subject.

Inaccurate eyewitness accounts are often at the root of wrongful convictions, and they are sometimes caused by what is known as "change blindness": the failure to spot a major change because something else draws your attention away from the spot where that change is taking place. Experiments have shown that test subjects often wrongly identify a person as a thief after seeing a video of a burglary simply because that person happened to appear in the clip. Imagine this scenario: you see Person 1 walk into a store and disappear behind a stack of boxes. When Person 2 appears from behind the boxes and steals something from the shop, it is not surprising when you subsequently identify Person 1 as the thief.

It is easy to do your own change blindness experiment. Take a photo of your living room and then remove one of the chairs. Take another photo from the exact same spot, upload the two photos to your computer, and open them one after the other with a white screen separating the photos for a brief interval. When you ask someone to view the photos, you will see that they will have great difficulty spotting any difference between them. However, when you let them look at the photos without the white screen in between, they will probably spot the difference straight away. Changes are also more noticeable when they happen in an interesting part of an image. For example, it is usually easy to see when someone has changed position in the middle of a photo. You can find lots of examples of change blindness on the Internet.

Change blindness also makes it more difficult to drive in the rain compared to driving when it is dry. In addition to the fact that the rain makes it harder to see other road users, the raindrops on your windshield and the windshield wipers cause visual disruptions that make it more difficult to react to changes in traffic conditions.

It is quite amazing the kinds of changes we sometimes manage to miss. In a particularly infamous experiment, test subjects were asked to go to a counter to pick up a form. The person behind the counter bends down to get the form and disappears from view for a moment. A different person then stands up in their place and hands the form to the test subject. Seventy-five percent of the test subjects failed to notice the switch. So don't feel too bad the next time your sweetheart walks into the room and says to you, "Well? What do you think?" and you have absolutely no idea which change in his or her appearance is being referred to...

Movie directors love change blindness. They try to create experiences that suck the viewer into their story. To do this, they need to ensure that each scene flows smoothly into the next so that the viewer does not experience any disruptive changes. The last thing they want is for the viewer to be aware of sudden changes in camera position, for example. They want viewers to experience their movie as if they were part of the action, but that feeling can be disrupted if the scene they are viewing changes abruptly. You wouldn't think it, but this actually happens quite a lot in movies. A typical Hollywood movie contains somewhere between one and two thousand edits, which means a new shot every three to five seconds or so.

There are a number of rules that have been devised by movie directors which they believe help to prevent changes from disrupting the viewer's experience and thus ensure a smooth viewing experience. These are known as continuity editing rules, and every single director and editor in the business knows them by heart. It is not easy to mask a change of camera position in a movie because it involves a complete change of image and not a minor alteration like those we have already seen in other examples in this chapter. Movie directors make clever use of the fact that viewers like to actively follow a story line and therefore do not look for the kinds of changes that result from using different camera positions. One rule is the so-called 180-degree rule, which states that shooting a scene with multiple cameras can only ever produce a pleasant and continuous viewing experience when the cameras are positioned at angles of 180 degrees. This allows objects in the scene to remain in the same position relative to each other. Imagine watching a scene in which the two main protagonists are having a conversation. It is no problem switching from one camera to the other as long as each character continues to occupy the same side of the frame. Any change to this setup will only confuse viewers and disrupt their viewing experience.

This also enables directors to zoom in during a scene. Many scenes begin with a wide shot before moving closer to the action, shot by shot. This literally draws the viewer into the movie, but the trick can only be used without becoming disruptive when there are no major changes in the composition of the scene. The viewer will be so intent on following the story line that he or she will not be able to focus any attention on such changes.

Very little research has been carried out into whether or not these rules actually stand up to scrutiny. In the few studies that have been done, test subjects were asked to make a note of every single change of camera position they noticed while watching a movie. The results showed that viewers tend to notice a change more often when it does not comply with the continuity editing rules. Changes that occur in the middle of a dialogue are less likely to be noticed because the viewer is too busy trying to follow the conversation and make sense of it in relation to the story line.

Other change blindness rules appear to apply to moviemaking, too. For example, changes are often less noticeable when they occur simultaneously with a major visual event, such as an explosion, just like a white screen that when shown very briefly can prevent test subjects from noticing differences between two seemingly identical images.

Another popular trick is the use of the actors' direction of view. For example, when an actor is looking at a relevant object that is located off camera, most viewers will be so busy trying to figure out what the actor is looking at that they will fail to notice when a change occurs in the scene, such as a car driving into the shot. This also makes it easy to switch between two actors who are looking at and talking to each other. We like to follow the direction of view of the person who is the focus of our attention because we assume that theirs is the most interesting point of view, so to speak.

So whether you are a radiologist, a security scanner operator, or a movie director, you still have to deal with the fact that we only ever have access to that small part of our world upon which we focus our attention. But how do we guide that attention? How do we decide which objects deserve further investigation? In the next chapter we will see that there is a continuous battle being fought for our attention between our internal world and our external world. *May I have your attention please?*

4   How to Find Your Tent at Glastonbury: Searching High and Low

Picture this: you are attending the Glastonbury Festival in England for the first time. You have been dancing your feet off all night, and the sun is already beginning to creep up over the horizon. You walk back to the campsite in the hope of catching some sleep. However, there is one problem: you have no idea where you pitched your tent. Of course, you knew that there would be 135,000 other visitors pitching their tents here, too, but you were confident that your memory would not fail you at this critical moment. Friends had advised you to download a handy app that you could use to pinpoint the location of your tent, but you didn't think you would need it. You now regret not heeding their advice. Fortunately, one thing you do remember is that your tent is green. You come up with a cunning plan that involves climbing up a lamppost so that you can get a better view and scanning the campsite for your tent.

You realize that you don't know where—not even approximately—on the campsite you pitched your tent, and so you don't even know where to start looking. If you had a vague idea of its location, you could at least focus your attention on that area. However, in the absence of such information you have no choice but to survey the entire campsite. One option would be to start searching from left to right and to check each tent individually. If there were 1,000 tents on the site and you needed one second to determine whether a particular tent was yours or not, the search would take you 1,000 seconds at the most. This would not be very efficient, however, because you already possess useful information about the color of your tent, presuming, of course, that the other tents on the campsite are not all the same color as yours.

Knowing that your tent is green, you can ignore all the other colors and search only for the green tents. But what strategy should you use? To answer this question, it might be useful to describe how we study visual search in the lab. In our experiments we ask test subjects to search for a particular object (target) in a visual environment containing other objects (distractors). Usually, the test subject is required to look for a single "anomalous" object and to press a button the moment he or she finds the unique object among the distractors. The reaction time is the time that elapses between the moment the search screen becomes visible and the moment the test subject pushes the button.

Take a look at the example in figure 4.1A, which is an image of a search screen in which the light-colored tent is the anomalous object and therefore the target. The image also contains three distractors. However, your attention is immediately captured by the light-colored tent. The fact that it has a different color results in a so-called "pop-out effect." The light-colored tent literally "pops out" from the screen. You may not have even started to look for the anomalous object, and yet your attention is immediately drawn to it. This is the reason why a green-colored Santa Claus is so easy to find in a group of red-colored helpers. You don't have to look for the anomalous Santa because it is immediately apparent where he is. It seems that our brain is programmed to move our attention instantly to information that deviates from the rest of the world around us. Advertisers often make good (and extremely irritating) use of this fact by displaying an anomalous moving object, like a dog, on advertising billboards during football matches.

Fortunately, not all objects are able to grab our attention so quickly. Otherwise, we would have great difficulty focusing on anything at all. But how do we know which objects cause a pop-out effect? To answer this question, I would like to invite you to sit in on one of the psychology lectures that I give to first-year students at the Utrecht University. Don't worry, you won't have to take an exam at the end, but it might be useful, in terms of everyday application, to find out what tricks we scientists use to reveal in experiments whether an object causes a pop-out effect or not. I will also need the information that is garnered from these experiments to be able to explain later on why faces are so special, why the attention of hungry people is automatically captured by pictures of food, and why spiders are often the focus of anxious people's attention.

So listen up, class. Imagine there is only one tent pitched at the campsite. In that case you will find your tent immediately without even looking for it. You might be tempted to conclude, therefore, that the time it takes you to find your tent is directly related to the number of tents on the camp-site. This is true in most situations, except in the case of a "pop-out." Look at the graph in figure 4.1B. The *y*-axis shows the test subject's reaction time, and the *x*-axis shows the number of distractors on the screen. Say we ask eight test subjects to conduct 30 searches in four different situations (with 8, 12, 16, and 20 distractors). We also insert 20 searches that do not contain a target object so that we can be sure the test subjects are actually search-ing for the object and not just pushing the button whenever something appears on the screen. We calculate their average reaction time and plot it on the graph.

In the case of a pop-out, the reaction time is not affected by the number of distractors. This means that the reaction time is the same in all four situ-ations. This can be seen in the example below in which the slope of the search function remains at 0 regardless of the number of distractors: no matter how many dark-colored tents there are, you will always spot your light-colored tent immediately. That's why, if you want to be able to find your children in the crowd on the Fourth of July, you shouldn't dress them up in red and blue but in green and orange instead. Unless, of course, every-one else has the same idea!

So why does the number of distractors make no difference? We have already seen how attention is required in order to gain access to the identity of an object. In the case of a pop-out, the visual system is able to detect a

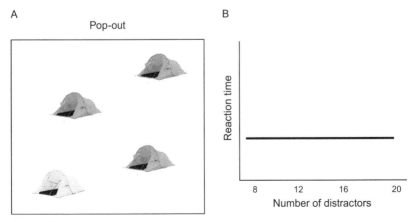

Figure 4.1
Pop-out searching

unique building block without any recourse to attention. Our brain does not know exactly what the object is initially, but it does know that there is "something" there with an anomalous shape or color. That is why our attention is drawn almost immediately to a unique object.

But back to your tent for a moment. Unfortunately, it does not result in a pop-out effect because it does not have a unique color. So what do the results of a search look like when there is no pop-out? In that case, the target object is not defined by a single anomalous building block, such as a unique color, but consists instead of a so-called conjunction of different building blocks. Look at the example in figure 4.2A: the target object is unique because it is the only *T* with a gray color—a unique conjunction of a shape and a color. There are other *T*s too, but they are all black. And there are other gray letters, but none of them are *T*s. So you have to inspect each object individually in order to find the target object. You focus your attention on one object, identify it as the target object or not, and continue your search until you have found your target. We call this serial searching, and in this case the slope of the search function is not 0 but maybe 50 milliseconds (see figure 4.2B). This means that with each additional distractor the reaction time increases by about 50 milliseconds.

So there is a difference between pop-out searching (or parallel searching) and serial searching. Unlike the serial process, pop-out searching appears to be an automatic process in which your attention is "spontaneously" drawn

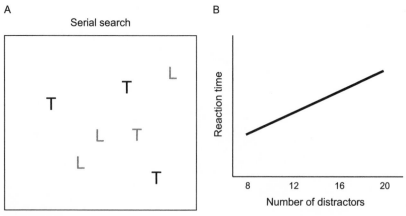

Figure 4.2
Serial searching.

to an object. But is this really an automatic process? This is a very important question because if it is, in fact, an automatic process, it means that
we do not have any control over our attention when there is a pop-out.
This may sound like we are all slaves to our attention systems, but we must
also appreciate just how important our reflexes are. In evolutionary terms,
it is pretty important that your attention is immediately grabbed by a lion
rushing toward you.

The kinds of searches that we have discussed up to this point cannot
tell us whether a pop-out draws our attention automatically or not because
we have also been looking for the anomalous object in our searches. Of
course, it is very welcome that the object grabs our attention because it
is exactly what we were trying to find. However, to find out whether the
pop-out is a reflex or not, we need to know what happens when the unique
object is a distractor and not a target. What happens when you are looking for your tent and suddenly an ambulance drives onto the campsite?
Does the ambulance automatically draw your attention? The answer to this
question is given in the second part of our lecture in which we discuss the
experiments carried out by my PhD supervisor, Prof. Jan Theeuwes, at the
VU University in Amsterdam. In his experiments, test subjects are asked to
find the anomalous shape among a group of other shapes (like the circle
among the triangles in figure 4.3). All of the shapes in the experiments are
white. However, half of the searches contain one unique distractor: in this
case a gray circle. This distractor has the same shape as the other distractors, but it also has a unique color that distinguishes it from all the other
shapes on the screen, including that of the target object. It is known that
we process information related to color faster than information related to
shape because color is a stronger building block than shape. In this case,
the distractor with the anomalous color is stronger than the object with the
unique shape. However, you are actively searching for the unique shape, so
doesn't that mean you should be able to ignore the distracting color?

The answer is no. If the unique distractor is not present, the unique
shape of the target object results in a pop-out effect, meaning that the
shape is able to grab your attention quite easily. However, it will take you
much longer to complete the search when the unique color is also present,
even though you are not actually looking for it. You simply cannot escape
its presence, not even when you know in advance that a distracting color
will be added to the picture or when you know which color will be doing

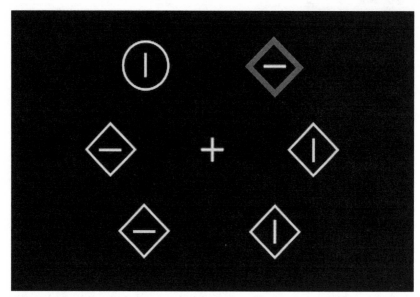

Figure 4.3
Capturing attention.

the distracting. In fact, you will not be able to ignore the distracting color
even if you repeat the search for hours on end. Unique color information
is so strong that it will always draw your attention. This means we can conclude that the drawing of our attention to a pop-out object is an automatic
process and one over which we have no control. We have a built-in reflex
that reacts automatically to new and unique information. It also explains
why stalking is such a successful hunting technique in the animal king-
dom. An animal that stalks its prey does not attract the victim's attention,
thus allowing it to creep closer without being noticed.

Making something suddenly appear is the best way of capturing a person's attention. Nothing activates our attention reflex more strongly than
a new object. It is easy to understand why we evolved this way. Objects
that appear abruptly can represent possible danger. Similarly, the sudden
disappearance of an object can also be a good reason for us to focus our
attention on it. When you can ensure that attention is not distracted by
another object, as in change blindness experiments, major changes, such as
the appearance of a new object or the sudden disappearance of an object,
will attract attention.

By now you will probably have realized that perceptibility and pop-out are closely related. A salient object often attracts our attention automatically. And, as is also the case with the perceptibility of an object, the extent to which an object automatically draws our attention is partly determined by the surrounding environment. It is by no means a given, therefore, that an object with a bright color will automatically grab our attention. At first it may seem like a good idea to design a brightly colored cover for your newly published book—like I tried to do—but that won't be much help on a shelf full of equally brightly colored books. A flashing banner on a website is surely a good way of attracting attention, but not when there are multiple flashing banners and video clips fighting for your attention, too. An effective advertisement is one that is adapted to suit its environment. A smart attention architect will develop an algorithm that can predict the visual characteristics of a website and automatically adjust their advertisement accordingly.

The above must be taken into consideration by anyone designing systems that require the user's attention to be drawn to the right place at the right time. Take, for instance, the cockpit of an airplane, which is fitted with lights whose aim is to attract the attention of the pilot. The thinking behind this is simple: a flashing light will ensure that the pilot's attention is drawn to an important piece of information. The problem, however, is that there can be a lot of flashing lights in a cockpit, with the result that pilots sometimes miss crucial visual information because of the overload. Studies have shown that in the case of cockpits and other similarly complicated environments, the context in which an attention-grabbing object is placed is crucial to its success. A green light that flashes when there is a change in the status of the automatic pilot can be easily overlooked when it is surrounded by lots of other green lights, but it will do its job perfectly when that is not the case.

In many situations we have very little control over which salient object will grab our attention, but there is another aspect of attention over which we can exercise some control: how quickly we are able to remove our attention from an object (also known as "disengagement"). When you have focused your attention on a particular spot in space, you can move it to a different location provided you first disengage your attention from the initial spot. The speed at which you can do that is often confused with how quickly something is able to grab your attention. For example, in the

Netherlands most people are familiar with the infamous billboard advertisements that once displayed pictures of female bums alongside one of the
country's main highways. The ads led to a lot of commotion because of the
fear that they would distract (male) drivers. This fear was probably well-
founded, but the potential distraction was not the result of this particular
advertisement's being able to attract attention more forcefully than, say, an
ad for a political party bearing nothing more suggestive than an election
slogan. Female bums do not contain any essential visual building blocks
that enable them to automatically attract attention: their color and shape
don't stand out any more than the colors and shapes of an election slogan,
for example. The distraction lies in the difficulty males experience in dis-
engaging from this kind of advertisement. Both advertisements will more
than likely attract the attention of drivers (picture yourself driving along
a highway on a dark night and seeing a brightly lit billboard), but it will
probably be more difficult for drivers to disengage their attention from the
advertisement with the bums once the female bums have been identified.
The visual system does not "see" any female bums until the driver's attention has actually shifted to the location.

Successful communication of a message depends upon a combination
of two factors: ensuring that your message attracts attention through visual
building blocks, and subsequently ensuring that the viewer's attention
remains fixed on the spot where your message is located. While it may be
possible to formulate universal rules for the first of these factors (and this
book is an attempt to do just that), the second factor involves an entirely
different question in the field of psychology: what do people find interesting? This is a question in which individual differences play a very significant role. After all, not everyone gets excited at the sight of a female
posterior. This is where market researchers and advertising agencies get to
prove their worth: they know exactly what is needed to attract and retain
the attention of the desired target group. This includes being aware of the
futility of using the female bum to attract the attention of a target group
that has no interest in such things. And for the readers among you who
are equally uninterested, let me reassure you: I will not be mentioning the
female bum again in this book.

Let's say you are afraid of spiders. Does that mean that spiders will automatically attract your attention? Over the past few years there has been an
explosion in the number of studies dedicated to establishing which objects

automatically attract our attention. It all began with a study into facial features, which demonstrated that faces enjoy a unique status and can even cause a pop-out effect, just like color and shape. It has been suggested that our attention is automatically drawn to a face because we see faces as socially relevant stimuli. But is that really so?

This is where the lessons learned in our previous lecture become useful. We have already seen that in order to establish whether an object can cause a pop-out effect or not, we have to vary the number of distractors to see what effect this has on the reaction times of test subjects who have been asked to search for a face. Studies like these have shown that faces can indeed cause a pop-out effect, which would seem to give us all good cause for celebration because it presents us with a terrific evolutionary tale that will undoubtedly prove very popular with the mainstream media (think of the headlines: "Faces are special!"). At long last, we attention scientists will have a good yarn to spin at parties instead of our usual diatribes about building blocks and colors and shapes and so forth.

However, we should be wary of celebrating too soon. After all, how can we be sure that a face is special just because of its social significance? Can't the pop-out be explained on the basis of shape and color alone? Imagine a situation in which a face is presented as a unique object surrounded by a number of gray squares. The face has a color but is also made up of different shapes, including a triangle (the nose) and two circles (the eyes). It could be the case that the face results in a pop-out effect simply because it boasts a number of unique visual features that the distractors do not possess. It seems it has nothing to do with its being an evolutionary object but instead can be explained on the basis of a unique combination of building blocks, nothing more and nothing less.

Fortunately, a number of ingenious ways have been devised to establish the validity or otherwise of the rather dull building-block explanation given above. One way is to provide the distractors with the same building blocks as the face, the only difference being in the composition of those blocks. So, if the mouth is at the top and the eyes are at the bottom, they do not form a face. However, this figure still has the same visual features as a face and also results in a pop-out effect. So maybe we can celebrate after all.

But what has any of this got to do with spiders? Spiders also attract our attention automatically, particularly when we are afraid of the little critters. This varies from person to person, unlike the way in which a unique color, shape, or face automatically grabs our attention. Only those of us who have a fear of spiders—arachnophobia—possess a special detection mechanism that can automatically spot a spider in our immediate environment without even having to shift our attention to the spider. If this is an example of a system that has been designed by evolution, it begs the question why spiders have this effect on some people but not on others. There is an interesting theory that attempts to explain why this varies from person to person: people who are anxious by nature constantly search their surroundings for threatening objects, even when there is no apparent need to do so. This is why they react quicker to a threatening object in a relatively safe environment, such as a laboratory. Their special detection mechanism is always "switched on," unlike less anxious people. It is possible that this kind of mechanism would have been very useful to humans way back in the distant past, but it is of little value in today's modern society. Indeed, some scientists believe that it may even have a detrimental effect. Fearful people, like those who suffer from anxiety disorders, are perpetually on their guard. They experience the world as a dangerous place and use a large portion of their attentional capacity to constantly monitor their surroundings. Their fear reduces their capacity to perform a task in a composed and focused manner, with all the resulting negative consequences.

Of course, a detection mechanism like the one above can be very useful in a genuinely threatening situation. When an object is linked to a painful experience, for example, we can be glad that it automatically attracts our attention. If someone who does not suffer from an anxiety disorder is shown a neutral object, like a red circle, that has been conditioned to produce an electric shock, then that person's attention will automatically be drawn to the red circle in all subsequent search experiments, even when the shock has not been administered for some time.

It appears that a pop-out effect can be caused not only by simple, anomalous building blocks, but also by extremely complex objects. For example, plates of food are known to automatically attract the attention of hungry people. Your visual system is set up to monitor the world around you within the framework of what you yourself regard as interesting—whether that is food when you are feeling hungry or spiders when you suffer from arachnophobia. We all know how difficult it is to concentrate when you are hungry. This is not only because internally you are focused on your hunger, but also because you will begin inspecting your surroundings in search of food. Personally speaking, whenever I am at a party I find it very hard to concentrate on what my conversation partner is saying if I am hungry and I know that there are plates of food being passed around the room.

Not all complex objects are able to automatically attract a person's attention. We know that a face can automatically draw attention, but this pop-out effect is the same for every other face as it is for your own. So your face does not attract any more attention than another face. There are differences, however, in how easy or difficult it is to disengage your attention: your own face is an interesting object and one to which your attention will tend to "stick." Of course, this can also be attributed to the surprise effect: we rarely see our own face anywhere other than in a mirror, and we would certainly not expect to see it in a laboratory experiment. I don't really see the evolutionary advantage of your face having a strong pop-out quality. After all, it's not like you are going to run into yourself regularly or that your doppelgänger is ever going to pose any real threat. Or maybe it's just a mechanism for automatically detecting mirrors so that we can avoid walking into them.

I'm getting sidetracked; time to return to Glastonbury. You are still hanging from a lamppost looking for your green tent, and you are desperate to get some sleep. Can your knowledge of the color of your tent help you to find it? The answer is yes. If you know that the object you are looking for is green, you can focus your attention on all the green objects and inspect each one to see if it is the one that you are trying to find. In that case, the search time will not be affected by the number of distractors *with a different color*. I should add that this is based on the assumption that the green color of the tent is different enough to distinguish it from the other tents. Your search will be less efficient if there are lots of tents whose color is close to the green of your tent on the color spectrum. So, to make this example work properly, we have to assume that all the different colors are easily distinguishable.

While searching the campsite, you move your attention voluntarily from one spot to the next in your visual world. By "voluntarily" we mean that you are free to choose how and when you move your attention. For example, you will do so when you know that something important is about to happen at a particular spot. Imagine that your friend calls you on the phone while you are swinging from the lamppost. He tells you that he is in the public washroom brushing his teeth and that he will be able to show you the way to your tent when he emerges from the building. You have already spotted the building in question, and so you move your attention to the door of the washroom. You keep your eyes focused on that spot so that you will see your friend when he comes out. That way you can be sure you won't miss him.

From experiments we know that we are able to process objects more quickly and more efficiently when they appear at the spot where our attention is focused. In one of these experiments, test subjects are asked to look at crosshairs, on either side of which two squares have been drawn. The test subjects are then told that a target object (e.g., a circle) will appear in one of the two squares and that their task is to press a button as quickly as possible when the circle appears. Just before the target object appears, an arrow is projected on the crosshairs. This arrow points either to the left or to the right and in 75 percent of the cases identifies the square in which the circle will appear. The arrow is of no help for the remaining 25 percent, in which the target object will appear in the opposite square. Test subjects could choose to ignore this information and focus their attention purely on the crosshairs. However, if they wish to perform well in the test because of the prospect of a reward or just because they want to get home quicker, they will use the arrow to move their attention to the square with the highest probability of the target object's appearing. The results of these experiments show that test subjects press the button faster when the target object appears in the appointed square than when it appears in the opposite one.

This kind of attentional shift is known as "cueing": your attention is shifted by information in the external world. A person looking for the restrooms in a building with which the person is familiar will follow the arrows that point the way. This is an example of what we call voluntary cueing, as the person could also have chosen to ignore the arrows (i.e., the "cue"). However, pop-out experiments have shown us that our attention can also be captured in an involuntary manner. If we want to study the automatic shift of attention, we need a cue that our test subjects will be unable to ignore. In the case of the example above, this can be achieved by briefly illuminating one of the squares just before the target object appears. The test subject's attention will automatically move to that square, just like it does with a pop-out. To ensure a purely automatic attentional shift, it is important that the cue does not reveal any information about where the target object is going to appear. Whereas, with the arrows, the cue provided the correct information about where the target object would appear 75 percent of the time, an automatic cue only does so 50 percent of the time. This means that we can be sure that the test subjects will not move their attention voluntarily on the basis of the cue. It is pointless to follow the cue voluntarily because it will not help them to perform the task.

Our attention system is constantly on the lookout for new information. It uses a mechanism that keeps track of those spots where our attention has already been focused but where no information of any importance was on offer. In other words, attention is a swift and impatient thing. For example, the pop-out effect of an automatic cue causes us to move our attention almost immediately, often within 100 milliseconds. That is as quick as a reflex action: something happens in the corner of your eye and you have to shift your attention as quickly as possible to find out what is going on. However, the rapid processing that occurs at the spot to which your attention has automatically been drawn does not last very long. If there are more than 200 milliseconds between the moment that the cue appears and the appearance of the target object, the rapid processing at the spot where the cue is located will already have ceased. In fact, your subsequent reactions to visual information at the spot where the cue is located will actually be slower. This is because your attention has already "moved on." This is very efficient, as there is little point in continuing to focus your attention on a spot where there is nothing to see. This spot becomes suppressed, and this is why it takes more time to shift our attention back to the spot where the cue is located.

The suppression of attention at a spot where no target object is to be found is very useful in our daily lives. If, for example, our attention is drawn by a flash of light but there is no important information for us at the point where the flash originated, then there is little point in turning our attention to that spot again. The best thing we can do is to ignore that spot. Suppressing a particular spot on the basis of an automatic attentional shift allows us to allocate our precious attention much more efficiently.

You might think that your attention can only be grabbed automatically by something that appears or changes abruptly in the corner of your eye. That is not the case. There are also situations in which our attention automatically moves in response to something that appears in the middle of a screen. It might be useful to pause for a moment and consider what that something could be: what kind of visual information presented to you in the middle of a screen would cause your attention to shift automatically (i.e., without any influence on your part) to a different location?

The most well-known example is that of a face with the eyes looking away, as shown in figure 4.4. Schematic faces are often used in experiments to ensure that gender and facial features do not play a role. When a face is looking to the left, your attention will automatically be drawn to the left. And even when the facial cue does not predict the spot where the target object will appear in an experiment, you will automatically move your attention in the direction the face is looking (see the Notes section). The explanation for this is simple: in evolutionary terms, it is sometimes important that we move our attention to the spot where someone else is looking. He or she may have detected something dangerous, and we will want to be able to react as quickly as possible to the threat.

Attention architects gladly make use of this kind of knowledge when designing websites and commercials. We know from the pop-out study that faces automatically capture our attention. If we combine this information with the knowledge that the attention of viewers always follows the direction in which a face is looking, then it can be very easy to manipulate their attention. If, for example, you want to draw a test subject's attention to a particular brand's logo, you can position a face so that it is looking in the direction of the logo. This will greatly increase the chances of someone moving their attention to that spot. However, the presence of faces can also have a negative effect on message delivery. In many TV programs, like the evening news, for example, faces are often shown in the background. They attract the attention of viewers, with the result that their focus shifts from the newsreader. And if the face in the background is looking away, it can lead to even more interference. It is therefore very important to choose the

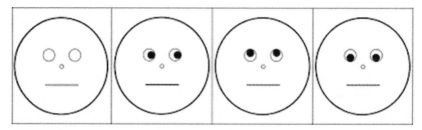

Figure 4.4
Facial cue.

right moment to show a face in the background, as well as ensuring that the face is looking in the desired direction.

This kind of attention is also known as "social attention," as it is said to possess a strong social component. You have to be able to relate to the other person in order to understand what his or her eye movements (to the left or right) mean. This idea is supported by the conclusion that a facial cue does not work with autistic children: they show no attentional shift in the direction of a facial cue. The explanation for this lies in the fact that these children are unable to relate to the other person, whereby they fail to understand that a sudden change in the direction in which that person is looking means that the person has spotted something important out of the corner of his or her eye. This is quite remarkable given that the expected effect is even found in three-month-old babies. The connection with autism may also explain why adult males are less responsive to facial cues than women, as research has shown that, generally speaking, males display more autistic traits than females.

Facial expressions of emotion also play a role in the following of facial cues. David Terburg and his colleagues at the University of Utrecht have carried out extensive research into the effect of emotions on facial cues. They have shown that test subjects are more inclined to follow the direction given by a face expressing fear than that given by a face expressing happiness. With a fearful face we experience a stronger inclination to move our attention in accordance with the facial cue because we assume that the other person's fear has something to do with whatever it is they are looking at. This is further evidence that the following of facial cues is an effect that has developed along evolutionary lines in relation to our ability to detect danger.

And that's not all. There is even a study that reveals the role played by political preference in determining the extent to which facial cues are followed. The idea is that people with a political preference for right-wing parties are more individualistic and, therefore, less susceptible to the influence of others, compared to people who have a preference for the left. The study was carried out in the United States among voters with conservative and liberal leanings. The results showed that facial cues have a much more significant effect on liberal-minded people than on their more conservative counterparts. It appears that individualistic thinking makes a person less inclined to follow others. Subsequent studies even demonstrated that test subjects automatically moved their attention to the left when shown an image of a "lefty," like Barack Obama. Viewing the face of a certain person can activate the association of that person with a location in space: when we see a lefty, we activate the left-hand side of our visual world. The fact that this happens automatically opens up interesting avenues for objective image studies.

These kinds of studies are as relevant as they are amusing. They tell us a lot about how we understand information. When test subjects are asked to read out words like "skyscraper" and "self-confidence," the way in which these words are processed causes their attention to literally shift upward. This shows that we understand these words by activating their spatial aspects. It is unclear whether or not this activation is required to actually understand certain words or whether it is just a side effect, but it does show just how closely our internal processing is related to the external world.

A good example of this kind of spatial processing is the effect of numerical cues. Seeing the number "1" automatically shifts a person's attention to the left while the opposite happens when the person sees the number "9." In experiments, the test subjects did not have to do anything with the number: it was simply displayed in the middle of a screen, after which a target object was shown on the left- or right-hand side. It appears that we associate the number "1" with "left." This is because of the way in which numbers are presented in the brain, that is, on a "mental number line." When we are asked to list a series of numbers, in the Western world we usually do this from left to right. I use the word "usually" very deliberately here because some people do this differently. For example, some people say that 1 is on top and 9 at the bottom (the numerical effect works vertically in their case). When we see a number, we immediately process it. This involves activating the spot that the number occupies on the mental number line, which results in our attention's being moved to the place in space that corresponds with the spot on the number line.

Numbers are a special case here because the attentional shift also occurs when there is no requirement to do anything with the number. For other categories that use spatial ranking, such as days of the week, months of the year, or letters of the alphabet, the attentional shift occurs only when test subjects are asked to perform a task using these words. For example, if you ask someone whether Monday is at the start or the end of the week, their attention will automatically shift to the left, although this does not happen when they are only asked to read the word.

Even the arrow in the original example of voluntary attentional shift above is not an entirely voluntary cue. We have all seen so many arrows during our lifetime that the presentation of an image of an arrow activates a slightly automatic attentional shift toward the arrow. In fact, it is very difficult to create a completely voluntary cue. Even an arbitrary object like a circle will result in an automatic attentional shift to the left if it has correctly indicated for a reasonably long period of time that a target object usually appears on the left-hand side. The brain is a learning system that continuously attempts to identify regularities in the world around us.

The moving of attention involves a complex interaction between the ability to detect possible threats in the external world and the need to successfully carry out our everyday activities. Our attention is constantly being pulled in one direction or the other, and the different spatial associations help us to understand the world around us. Searching for your tent at Glastonbury is by no means an easy task, but the more information you have about the color, shape, and location of your tent, the quicker you will find it. The fastest way of finding it is for something to happen next to your tent that will automatically attract your attention. A flashing light or a flag attached to the tent will help to draw your attention to the right spot. In the absence of these items, however, you could be in for a long night.

5 Your Gateway to the Visual World: How Your Eyes Betray Your Thoughts

The Dutch government recently launched a safe-driving campaign with a spot on TV that shows a man driving a car in an urban area. We see him bobbing his head up and down to the rhythm of a reggae song playing on his car radio. Every time his head bobs down, he looks at the speedometer, and when it bobs up, he has his eyes on the road. We then see him braking to allow a pedestrian to cross. The voice-over compliments him on his driving: "Very good! The more often you check your speed within the city limits, the less likely you are to drive too fast."

The aim of the campaign is to encourage drivers to check their speed more often when driving in built-up areas. The information accompanying the campaign reveals that at least ten pedestrians and cyclists are killed each year as a result of cars driving up to 15 kilometers per hour (about 9 mph) faster than the speed limits in urban areas. Another 200 people suffer serious injuries that have to be treated in a hospital. As many of these accidents are the result of drivers' not knowing how fast they are driving, the government is doing all it can to make them more aware of their speed on the road.

The campaign's website explains that a car traveling at 35 kilometers per hour (just under 22 mph) in a 30-kilometer (about 18-mile) zone needs an extra 3.2 meters (about 10 feet) to come to a halt after applying the brakes. However, if you check your speedometer regularly and adjust your speed accordingly, you won't need this extra braking distance. Although the campaign is to be applauded, it fails to take one important factor into consideration: the time it takes you to check your speedometer. We know from previous chapters that you cannot process the information about your speed without moving your eyes toward the speedometer; it can't be done out of the corner of your eye. So in order to check your speed regularly, you have to move your eyes from the road to the speedometer very often. This begs the question: is this safe? Given the fact that we are unable to register any visual information when moving our eyes, we are literally blind to the outside world when we do so. And not only does checking your speedometer require you to move your eyes, but it also costs you precious time to process the information on the device. And while you are busy doing that, you run a greater risk of missing important information outside on the road.

Okay, let's do the math: checking your speed requires you to move your eyes to the speedometer, take in the information, and then move your eyes back to the road again. We know that it takes at least 120 milliseconds to make a relatively large eye movement, and you need to make two such movements: one to the speedometer and one back to the road. We also have to add the time needed to process the information on the speedometer: around 110 milliseconds. So the entire process takes around 350 milliseconds, which translates into 2.9 meters (about 9.5 feet) at a speed of 30 kilometers per hour and 3.4 meters (about 11 feet) at 35 kilometers per hour. If someone crosses the road at the moment when you start to move your eyes toward the speedometer, you will need an extra 2.9 meters' braking distance even when you are adhering to the speed limit.

The government's safety campaign could quite easily backfire altogether. If the information provided causes drivers who normally do not exceed the speed limit to check their speedometer more often, it could lead to situations that are even more dangerous than when someone is driving too fast. It would appear to be a lot safer for drivers to keep their eyes on the road and to look at their speedometer as little as possible. This is a good example of a situation in which auditory information would be much more useful. When we are driving a car, our visual system is far too busy paying attention to what other road users are doing, and so it would be more efficient to use one of our other senses to process other important information. We are perfectly capable of picking up and interpreting an auditory signal while our eyes are busy looking at something else. This is different, however, when we are engaged in a telephone conversation because that requires us to think, too. A simple auditory signal that is activated when we exceed the speed limit can be enough to make us slow down and at no expense to the visual process.

It may seem a little odd to start a chapter about eye movements with a story about a situation in which they represent a problem. Eye movements are primarily a way of solving the lack of visual acuity in the corners of our eyes. We make thousands of eye movements each day in order to focus the sharpest part of the retina on objects in the external visual world. Six muscles are tasked with making this happen at an incredible speed. Eye movements are among the fastest movements that humans can make. This speed is crucial because visual information can become relevant in an instant. For example, we will want to know as quickly as possible whether the blurry moving object on the side of the road is a child about to cross over or just a flag fluttering in the wind. And the speed is also necessary because we are unable to process visual information when moving our eyes.

The magnitude of the eye movement and the length of time that the eye remains still depend on the task being carried out. When we are reading, we usually make small eye movements (approximately eight to nine letters in length), but this can double in magnitude when we are looking at a picture. A complicated search requires us to keep our eyes still for longer, like when we are trying to find our bin among all the other bins out on the street on garbage collection day. In this scenario we may even end up fixating our eyes on our bin several times in succession but not for long enough, resulting in our failure to identify it.

Even though we perform an eye movement every three seconds, no one ever complains of being exhausted as a result. We perform the task without noticing it and without any effort. We do have some control over how and where we move our eyes, but the majority of our eye movements are reflex actions. This can be deduced from the reaction time involved. In some cases it takes as little as 100 milliseconds for our eyes to move in the direction of a flash of light. This is far too short for anyone to be able to make a conscious decision based on objectives or intentions. You can do the math yourself: from the moment that the flash of light falls on your retina, it takes about 50 milliseconds for that information to reach the visual system via the optic nerve. If you subtract 30 milliseconds for the required muscle movement, you are left with 20 milliseconds in which to make a decision. This kind of eye movement is an uncontrollable reflex that is executed in the primitive regions of the brain. The parts of the brain that are responsible for more complex skills, like setting personal objectives, are far too slow to be able to influence this process.

Although we have two eyes, we are only able to fixate our gaze on one point in space at a time. This can lead to problems in many kinds of situations, including the romantic kind. The first time I found myself attempting to look deep into a girl's eyes, I quickly discovered that it simply wasn't possible. We cannot look into both eyes at once, and so we always have to choose one eye or the other. The only way to even get close to looking at both eyes at the same time is by switching our gaze rapidly between them. Unfortunately, the corners of our eyes do not have the kind of sharp focus that would allow us to concentrate our gaze on a spot between the eyes and still be able to see both eyes clearly. "I get butterflies in my tummy every time I look into your eye…" may not sound very romantic, but it is scientifically accurate.

The continuous movement of the eyes presents our visual system with all kinds of interesting problems. If we were to see the actual images that fall on our retina, they would appear fuzzy and shaky because of the movements that the eyes make. Each individual eye movement is responsible for a separate part of the visual world falling on the retina. And yet we do not experience our visual world as a series of continuously shifting images, but rather as constant and fluid. We never become disoriented as a result of moving our eyes.

There is a difference between a retinal representation (the image that falls on the retina) and a spatiotopic representation (where an object is located in relation to your body). Consider the example in figure 5.1. In the image on the left, the fork is next to the man's right hand. He is looking at a spot to the left of the fork, and so the fork is visible in his right visual field.

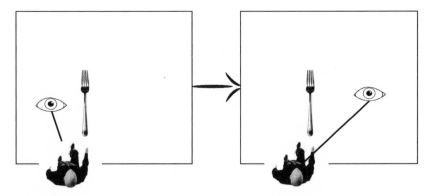

Figure 5.1
Eye movement.

When he moves his eyes to a different spot in space, the image that falls on his retina changes. However, his image of the world around him remains stable. If he moves his eyes to the right, the fork will appear in his left visual field, but he will know that it is still the same fork as before, despite the change. Although the (spatiotopic) location of the fork has not changed as a result of the eye movement, the retinal presentation (where the image of the fork falls on the retina) has.

We usually know exactly where important objects are located after each eye movement. We update the visual representations of an object after each movement so that a fork, for example, is still the same object after each subsequent eye movement. To facilitate this, the visual system needs to know the magnitude of each eye movement. Whenever a movement of a certain magnitude is made, an update of the same magnitude must be carried out as well. Various studies of monkeys' brains have shown that an update of the same magnitude takes place just before an eye movement. The brain seems to possess a kind of map that helps to ensure that we know exactly where important objects are located in relation to ourselves, regardless of where we happen to be looking.

Of course, we do not update the representation of every object around us after each eye movement. It would cost far too much time and energy to carry out such complicated calculations. We do this only for the most important objects—a maximum of three or four. This is primarily because we use the visual world as a kind of external hard drive. Given that we can only represent a few objects in our internal world at any given moment, it is only these objects that we take into consideration when we move our eyes. For example, it is impossible for us to know the exact location of each knife on a table set for eight people after each eye movement that we make.

A similar updating process occurs in the case of memory signals. In order to know where you have already searched when you are looking for something, you need to remember those locations. If you don't do this, you will only end up searching the same places over and over again. We have already seen that we suppress attention to prevent us from searching the same place twice. With each eye movement, the locations that have already been searched are projected onto a different part of the retina. So if you want to remember where you have already searched, you have to update these parts, too. This process has its limits, however, as you are only capable of recalling a handful of places you have previously searched. One solution is to add some structure to your search. If you know that you always search your bookcase, for example, from top left to bottom right, then all you need to do is to remember this strategy. This is why searching in a haphazard manner is so inefficient: you are never able to recall all of the individual locations that you have already searched.

Another limit of the eye-movement system is the number of movements that can be carried out at any given time: only one. This means that the decision-making process that determines the direction of each eye movement is constant and continuous. The process is very similar to the competition between the internal and external worlds in the battle for our attention. Visual information in the external world can force us to make an eye movement on the basis of that information. If we see a flash of light out of the corner of our eye, we will automatically make an eye movement in its direction. However, there are also many situations in which we are able to suppress this automatic eye movement. The most basic experiment used to demonstrate this is the antisaccade task. In this experiment, test subjects are asked to look at a fixed point in the middle of a computer screen. The rest of the screen is empty. After a short interval, an object (e.g., a ball) appears on the right- or left-hand side of the fixed point. There are two possible tasks in this experiment, and the color of the fixed point tells the test subject which task to perform. When the fixed point is green, the test subject must make an eye movement as quickly as possible in the direction of the ball. This generally does not present any problems.

However, a problem does arise when the fixed point turns red. In this task, the test subject is required to make an eye movement to the other side, that is, in the opposite direction to where the ball appears (a so-called antisaccade). Test subjects often get this wrong by making an eye movement toward the ball instead. It is important to know that no visual information at all is displayed on the opposite side. The test subject is being asked, therefore, to make an eye movement toward an empty spot. This eye movement is directed entirely by the internal world—the instructions for the task in this case. Generally speaking, we are pretty good at executing this kind of eye movement, except in this situation where there is competition between the voluntary eye movement and the reflexive movement in the direction of the ball. The ball is a very salient object because it is a new object that appears at a spot where no visual information was previously present. This will result in a strong attentional shift toward the ball, which in turn elicits a reflexive eye movement in the direction of the ball. And given that we are only capable of making one eye movement at a time, the competition between the two kinds of movements has to be resolved. The movement toward the ball must be inhibited, and to do this successfully, we need to be able to exercise a certain amount of control over our eye movements. However, this is not always possible. A young and healthy test subject will get it wrong around 15 percent of the time when performing this antisaccade task. In those cases, the eye movement toward the ball will not be inhibited sufficiently and the reflexive eye movement will emerge as the winner.

The reflexive character of these mistakes is also evident in the eye's reaction time. Whenever a mistake is made, the reaction time for the eye movement is often so short that we assume that it has to be a reflex action. On the other hand, the reaction time is considerably longer when a correct eye movement is made toward the empty spot. Inhibiting a reflexive eye movement takes time, and movements that are carried out on the basis of cognitive factors, such as instructions for a task, are initiated more slowly, as the cognitive control systems in our brain are a lot slower than our reflexive systems.

The cognitive control systems responsible for performing voluntary eye movements are situated primarily in the frontal cortex. Patients who have suffered damage to this region are known to make a greater number of mistakes when performing the antisaccade task. Elderly people and young children also make more mistakes in this task because they have a lower level of cognitive control. This is why the antisaccade task is often used to establish the level of cognitive control of patients who suffer from mental disorders. It is, after all, a relatively simple task, the results of which can be easily translated into levels of cognitive control. For example, children with attention deficit hyperactivity disorder (ADHD) make more mistakes during this task than children of the same age who do not suffer from the disorder. They tend to be driven more by their impulses, and this manifests itself in the form of reflexive eye movements. This also tells us something about how these children experience their visual world: when they are asked to perform a task, they are more likely to be distracted by an unexpected event in the visual external world.

Training can have a positive effect on people's ability to restrict unwanted eye movements. Test subjects who were given a few days' training in how to perform the antisaccade task became less prone to error. It is unclear, however, whether the improvement shown also leads to an improvement in the ability to control one's eye movements in normal, everyday situations. Experiments like this often result in improvements in areas that have little to do with life outside the lab. For example, you might get better at remembering whether you need to perform a normal saccade or an antisaccade, but that knowledge won't be of much use to you in the outside world. One experiment involved an attempt to find out whether gamers perform better than average at the antisaccade task. They don't, as it turns out. Professional athletes were also shown to make just as many mistakes as normal test subjects. Although they appear to have a shorter reaction time for the antisaccade task, they do not possess a greater level of control over their reflexes.

One possible way of achieving more control over your reflexes is by creating situations in which your body produces more dopamine. Dopamine plays an important role in strengthening the cognitive control you can exercise over your reflexes. If you get a group of test subjects to watch a funny movie, they will subsequently perform better at the antisaccade task. The idea behind this is that a positive frame of mind boosts the production of dopamine in the brain, which leads to greater cognitive control. This hypothesis appears to be correct. Studies have shown that schizophrenics make more mistakes when performing the antisaccade task, and schizophrenia is known to arise because of problems with the balance of dopamine in the brain.

Of course, the best way to stop making the wrong eye movements is not to make any eye movements at all. This might seem like a frivolous remark, but studies have shown that there is an important difference between attention and eye movements. It is often suggested that you can pinpoint the focus of people's attention by following their eye movements. After all, their attention must be focused on whatever it is they are looking at. This is not necessarily so. People are quite capable of shifting their attention without making any eye movements at all. This can come in very handy when you find yourself talking to someone at a party but are actually more interested in someone else across the room. You will be able to continue to look your conversation partner in the eye while at the same time focusing your attention on the other person. A funny evolutionary explanation for this human quality can be found in the hierarchy that prevails among gorillas, where you will find yourself in serious trouble if you look the alpha male directly in the eye. However, although a gorilla that is lower in rank will avoid eye contact with the alpha male, he will still want to be able to keep an eye on him, so to speak. Similarly, in the human world it is sometimes wiser not to look someone in the eye when they pose a threat to your safety. This can help prevent possible escalation (at least I hope it does the next time I run into a stranger in a dark alleyway).

If I were to announce that you will hear a loud explosion ten seconds from now and that it is important not to look in the direction of the sound, you will be able to keep your eyes still. Reflexive eye movements only occur when test subjects find themselves in a situation where they are required to make eye movements. If no eye movement is made in the direction of a salient object, this does not necessarily mean that no attention has been paid to the object. Drawing attention and eliciting an eye movement are two different things. When a pop-out object appears on a screen during a search assignment, your gaze is not always attracted to that spot. The reaction time for the search does slow down, however. The reason that extra time is required to carry out your search is that your attention has to move to the salient object. And although an eye movement toward the salient object is programmed to occur, it does not take place. The movement is inhibited, but the time that it takes to do this contributes to the slower reaction time for the search.

There is a lot of evidence to suggest that attention and eye movements are parts of the same system. The regions in the brain that are responsible for these two functions overlap to a significant degree, and shifts in attention and eye movements co-occur in many situations. For example, while it is possible to refrain from making an eye movement toward an attention-seeking object, this does not apply the other way around: you cannot make an eye movement without your attention first going to the end point of that movement. Attention can be said to precede eye movement. You will react quicker to an object when it appears at the spot to where the eye movement travels. This probably sounds quite logical. Moving attention takes less time to complete than the more sluggish movement of the eye, which also requires the use of the eye muscles. But because eye movements always travel to the most important spot in the visual world, it is at its most efficient when visual attention precedes it in the same direction. This always makes me think of the speedy, little red car that travels ahead of the larger, more cumbersome fire truck when responding to an emergency.

Okay, that was all rather theoretical, but what happens when you put this into practice? If, as an advertiser, you want people to be able to identify your logo, you will need them to make an eye movement toward that logo. In the case of a small logo, a shift in attention is not sufficient to fully facilitate identification because of the limited visual acuity in the corner of the eye. A useful trick for ensuring that a person's eyes are drawn to a logo is to locate it along the line of an advertisement's logical direction of view. As an example, my colleague, Ignace Hooge, often refers to the Benetton advertisement in which the logo is located at the end of a short line of text. The text itself is so salient that the viewer cannot help but read it. Given that the logo is located to the right of the text and is itself also very salient, the viewer's eyes will more than likely take in the logo as well. This is a good example of a visually effective advertisement. It is one thing to get people to look at your advertisement, but if they don't remember the name of your company, then it will not have had the desired effect. An advertisement that shows a large group of people staring straight at the viewer and in which the logo is very small and tucked away in a corner is unlikely to be very effective. Viewers will spend so much time moving their eyes from one face to the next that they might not even get around to noticing the logo.

The same Ignace Hooge has been using this kind of information for many years to advise businesses on the visual effectiveness of their advertisements. It is possible to measure the eye movements of consumers to determine what they have seen. An advertisement in which consumers do not look at the logo or the name of the company in question will not achieve the goal that the attention architect has in mind. The location of the advertisement also plays a role. When a page in a magazine contains information that, for whatever reason, is visually very attractive, the reader is less likely to make eye movements to the advertisement (even if the logo has been positioned very cleverly). In any case, not many people buy magazines to look at the ads. So, as an attention architect, you have a very small window of opportunity in which to elicit eye movements. An advertisement is unlikely to be effective when there is a lot of other visually attractive information competing for attention. A good, full-color advertisement in the stocks and shares section might be the most effective option, although in that case you run the risk of the reader's skipping the page altogether.

The location of an advertisement determines to a large extent whether or not the viewer's eyes will be drawn to it. Our previous experiences also play a significant role. We usually know where we can expect to find the ads in a magazine or newspaper and also what they look like. Most of us do not read a magazine for the ads it contains and will try to ignore the places where they are likely to appear. I am always unpleasantly surprised every time I find an advertisement taking up a whole right-hand page in a newspaper because I never expect to see an ad there. My eyes will automatically fall on the advertisement. This is why the cost of placing an ad on the right-hand page is a lot higher than placing one on the left. An advertisement on the right is far more visually effective.

We are usually able to recognize an advertisement or a banner on a website out of the corner of our eye, as they often have a distinctive design. On the one hand, this can help to attract our attention, but on the other hand, because we know from its design that we are dealing with an advertisement, we are also able to disengage our attention very quickly and so prevent our eyes from being drawn to the ad. Thanks to our previous experience with advertisements and banners, one of our main tasks when looking at a magazine or a website is to avoid the ads. From the advertiser's point of view, it would be better to design advertisements that look more like the information on the page on which they are printed. Sometimes I find myself reading an article only to stop halfway when I realize that it is actually an ad. Whenever an advertisement is almost indistinguishable from the normal articles in a newspaper, the paper's complaints desk usually gets very busy.

The above may be a little confusing. What is the most effective method: creating an advertisement that stands out like a sore thumb or one that blends in perfectly with its surroundings? The answer lies in the size of the attentional spotlight. When we visit a website with which we are familiar, we make our spotlight small and focus only on those parts of the website that we recognize as actually belonging to the website. A very striking advertisement located on the side of the page will have little impact in this case. Advertisements that fall outside the spotlight do not attract attention. It would be wiser to place a less salient ad on this kind of website because that would increase the chances of the viewer's regarding the advertisement as a genuine part of the site. The situation is different in the case of a website where we do not have a clear-cut task to perform or are searching for something without knowing exactly what it looks like. Our spotlight is then larger, and our attention is not guided by our knowledge of the characteristics of the object we are searching for. In that case, a nonsalient advertisement is useless because we will not confuse it with the content of the rest of the website due to our lack of knowledge about the site. A salient advertisement would be much more effective.

Websites are often prone to "banner blindness": users do not look at the ads despite all the flashing and flickering. In fact, all that flashing actually makes us better at suppressing our eye movements toward an ad. Pop-up screens are similarly ineffective: we know exactly what to expect and immediately close all pop-ups without even looking at their content. Using a small spotlight, we guide our mouse toward the little x as soon as the pop-up appears. Measurements of the eye movements of people visiting online news sites have shown that users make significantly more eye movements toward advertisements that are placed between news items than those located on the sidebar. The effectiveness of an ad is therefore directly related to the task being performed by the user: if the task requires a large spotlight, a pop-out ad will work well, but it is better to use a less salient ad when the spotlight being used is small.

To perform a task efficiently, it is very important that your eye movements are made in the right direction. So can you train yourself to do this? We have already seen that studying medical scans is a very complicated task in which it is crucial that the diagnostician's eyes be focused on the right spots. Experts are good at developing strategies for their work, but they are often unaware of how they do what they do. It can be compared to riding a bicycle: it is almost impossible to explain to a child how to ride a bike. It just doesn't involve conscious competencies.

By way of instruction, you could show a student what an expert looks at when he or she is scrutinizing a scan. This appears to be very effective, and not only for radiologists but also for all kinds of people who perform complicated tasks where looking in the right places is required. Take, for example, the practice of inspecting an airplane for mechanical faults before it embarks on its next flight. Part of this job involves a visual check of the exterior of the airplane. When students are shown a video in which the most effective eye movements are projected on the screen, they learn faster and more effectively how to perform this kind of inspection.

Demonstrating the eye movements of experts can also help when trying to solve a difficult puzzle or when learning how to drive. Everyone who has had driving lessons knows just how much emphasis is placed on looking at the right places. You have to look in your wing mirrors and your rearview mirror while at the same time keeping your eyes on the road. Your examiner will watch your eye movements very closely. When I failed my first driving test, it was because I hadn't looked at the right places at the most critical moments. I tried to tell my examiner that there is a difference between where you look and where your attention is focused, but he was having none of it. I still believe that I have enough visual resolution in the corner of my eye to be able to know when a cyclist is approaching just by moving my attention, while at the same time keeping my eyes on the middle of the road. Okay, I might not be able to tell the color of the cyclist's eyes, but what does that matter? That said, you could be forgiven for thinking that the main reason I wrote this book was to be able to explain all of this to my driving test examiner...

But back to the science. When, after taking their first three driving lessons, a group of new test subjects was shown a video in which the eye movements of an experienced driver were projected on the screen in the form of a moving ball, they subsequently made larger eye movements than the group that had not been shown the video. They also looked more often and longer at the right places, like the mirrors, when driving. Even after six months, the effect of this intervention was still apparent in the experimental group.

There is a new technique that uses the eye movements of experts to directly influence students. It uses subtle changes in the corner of your eye that cause your eyes to be drawn to the place where the change is taking place. The technique is called "subtle gaze manipulation." It can involve something like a brief change in color that disappears as soon as your eye has been drawn to the spot in question. You are not aware of the change, but you still allow it to guide you. This technique makes it possible to guide a student's gaze toward a place where an expert would look if he or she were performing the task. The expert serves as a model, and the aim is to try to get the student to act as the expert would. The idea is that by the end of the training the student will have acquired the same behavior as the expert. This training has been seen to have a positive effect on the ability to spot malignant tissue in a mammography.

Given all these new discoveries, it is no wonder that the future seems bright for applications that make use of eye tracking. Devices that measure eye movements—so-called eye trackers—are becoming more and more advanced. Eye trackers use infrared cameras to monitor the movements of the eyes. They are quickly becoming cheaper and smaller: ten years ago they required a hefty investment—a good eye tracker would set you back a few thousand dollars. Now, however, a reasonably good one will cost you no more than $200 or so. At that price it might not be accurate right down to the last millimeter, but that is not always necessary. If you only need to know which object someone is looking at, a margin of error of one centimeter on a computer screen is usually not a major problem.

Thanks to their falling cost and the many possibilities they offer, it is expected that eye trackers will soon become a feature of communication tools like mobile phones, tablets, and laptops. When you know what a user likes to look at, it is very easy to offer that user information in a way that you know he or she will notice. And eye trackers will also tell you which information the user has not looked at. Imagine a car that has been fitted with an eye tracker. It will be able to warn a driver if the driver takes his or her eyes off the road. And a system like this could even be programmed to wake up a driver who has fallen asleep at the wheel.

A person's eye movements can even reveal what they are doing. The eye movements of a person who is reading are very different from those of someone who is searching for something. Think back to the description of the ideal advertisement that was given earlier. If advertisers can establish whether a consumer tends to use a small or a large spotlight on the basis of that consumer's eye movements, they can adjust their ads accordingly.

Inexpensive eye trackers also offer the possibility of operating computers through eye movements. Imagine you are baking a cake and your fingers have become all sticky. A system that would allow you to navigate a web page using your eyes instead of your hands would no doubt come in very handy; blink twice for double-click, and so on. This kind of system would not only be useful when you have sticky fingers, of course. It could also be of great benefit to people who are no longer able to operate a mouse or touch a screen.

When we have to activate a system using a button, there is always a delay between the time it takes you to decide to press the button and the act of pressing it. A built-in eye tracker can allow a system to prepare for a certain task before the button is actually pressed. Let's say it takes 500 milliseconds for an airplane to start its descent after the button for the landing procedure has been pressed. The pilot's eyes will already be on the button before it is pressed 200 milliseconds later. Based on the pilot's eye movements, it would be possible to shorten the 500-millisecond delay by activating the landing procedure before the pilot actually presses the button. Of course, it would also have to be possible to abort the procedure should the pilot decide not to press the button, but such a system could certainly help to save time.

The first tablets with eye trackers are already being introduced to the market. In 2015, Apple was granted a patent for a system that will enable users to move the cursor with their eyes. And it probably won't be long before Apple starts using the high-quality cameras in its telephones and tablets to monitor our eye movements. Why use your finger to turn the page on your e-book when an eye tracker can do it for you automatically after you have reached the end of the page?

Eye trackers can be extremely useful reading aids. Research has shown that people who experience difficulty with reading have significantly different eye-movement patterns compared to people who have no such problems. They tend to jump back to words they have already read and focus on the wrong parts of words, and they also display longer fixation periods (the length of time that the eye remains fixed on a word). Although eye movements are still not being used much in the diagnosis of reading problems, the availability of inexpensive eye trackers is sure to accelerate their introduction into clinical practice. Many of those who suffer reading problems find it difficult to maintain a good pace when reading because they cannot "anchor" their eyes to the words. Eye trackers may offer a solution. We (Westerners) read horizontally and from left to right, but when we reach the end of a line of text, we have to go all the way back to the left to continue reading. A less able reader will find it difficult to make the required eye movement to the correct position. An eye tracker could help by monitoring the reader's eye movements and activating a brief flash of light at the start of the next line when the reader reaches the end of the previous one. The eyes are drawn automatically to the correct spot, thus improving the person's reading ability. Monitoring the eye movements of good readers can help less able readers to determine which parts of a sentence they should look at to make their reading ability more efficient.

Eye movements can also be used to detect neurological disorders. Studies have shown that certain eye-movement patterns are characteristic of disorders like ADHD, Parkinson's disease, and fetal alcohol spectrum disorder, a condition that causes mental problems (such as poor memory) among children whose mothers consume excessive amounts of alcohol during pregnancy. Researchers use the eye movements made during a 20-minute period of watching television as a biomarker for these kinds of disorders. As with saliva and neuropsychological tests or MRI (magnetic resonance imaging) scans, an eye-movement pattern also possesses a biometric signature that can be identified using an algorithm. The advantage of tracking eye movements, compared to other biomarkers, is that it is both easy and inexpensive. Watching television is a task that doesn't take much explaining and is less likely to be misinterpreted, unlike many other behavioral activities. This is primarily of benefit to young children and the elderly. The less you need to explain, the better. A computational model of visual attention was able to identify 224 quantitative characteristics from eye movements, which smart algorithms then used to determine the critical characteristics of certain disorders.

In the future we will use eye movements a lot more often when interacting with our surroundings, as well as in the diagnosis of neurological disorders and the treatment of related problems. Eye movements are very suitable because they represent our gateway to the visual world. If we want to know how someone experiences the world, all we need to do is look at how the person moves his or her eyes. A child with ADHD makes very different eye movements than a child who does not suffer from the condition. All kinds of higher-order cognitive problems can influence a person's decision on where to move his or her eyes. That decision also has a bearing on subsequent decisions because those decisions are based on the visual information that the brain takes on board. So while our cognition can be said to determine our eye movements, the reverse also holds true.

The guidance of our eye movements is part of a fantastic system, one that has a lot to teach us. Indeed, the system often seems like a metaphor for our "real" lives. The decision to make an eye movement toward a certain point happens very fast, but it can also be corrected rapidly if so required. At the moment that the eye begins to move, all that is known is the initial direction; the precise end point is determined during the course of the eye movement. The eye-movement system monitors the difference between the desired end point of the eye movement and its initial direction. So if the eye movement ends up somewhere other than where it was intended to go—for example, because of the interference of a distractor—a new eye movement can be programmed extremely quickly. The distractor is fixated only very briefly, and the eyes can move on to the correct location almost immediately. The eye-movement system does not deliberate extensively, therefore, on its decisions but focuses instead on the fastest possible course of action, one in which errors can be corrected along the way. Just like we often do in real life.

6 The Influence of the Past on Your Attention in the Present: You Only See What You Expect to See

On a quiet Wednesday morning in a small village in the Netherlands, a car is making its way down a road through town. The driver indicates that he intends to turn left onto a side street. He brakes and checks to see if there is any oncoming traffic. Seeing no other vehicles, he puts his foot on the gas to make the turn. All of a sudden there is a tremendous bang when a scooter collides with his car. Fortunately, the man riding the scooter doesn't suffer any serious injuries, but he does end up taking the driver to court. The judge finds the driver of the car guilty of reckless driving. If he had been paying attention, he would have seen the scooter and not ended up causing an accident. The driver disagrees: he had been paying attention; he just didn't see the scooter coming. In his opinion, the fact that he had stopped before turning proves that he had been driving carefully. The driver of the scooter also confirms that the car had stopped before attempting to turn into the street.

How did the driver not the notice the oncoming scooter, and should he have to take all the blame? It happened during the day, the driver was not in a hurry, and he had never had an accident before. Let's analyze the situation a little more closely. The scooter was driving at 45 kilometers per hour (about 28 mph) on the road, in full compliance with Dutch traffic regulations that stipulate that scooters must drive on the public road in built-up areas and not on the bicycle path.

If we look more closely at the circumstances, we see that the road is not the kind that would normally be found in a built-up area. The bicycle paths are separated from the road by a narrow border of grass and trees. It looks more like a country road; there are no houses around. It doesn't feel like you're driving through a town. It is not a road where you would expect there to be a 50-kilometers-per-hour (about 31-mph) speed limit or where you would expect to see a scooter, as they normally drive on the bicycle paths. So it is possible that the driver of the car did look to see if there was any oncoming traffic but did not expect that traffic to include a scooter. His target object was a larger vehicle, like a car or a bus. He may have overlooked the scooter simply because it did not match his search criteria.

Expectations are crucial when it comes to noticing objects. Road users with the wrong expectations may simply be unable to detect other oncoming traffic. Experiments have shown that road users can even fail to see a clearly visible police car when it is parked in an unexpected spot, like on the hard shoulder. The same can apply in a situation where you have a clearly visible cyclist and a separate bicycle path next to the road. A driver of a car will fail to see the cyclist more often when he or she cycles on the road instead of on the bicycle path. Of course, this book is not about the justice system, and it is a strictly legal matter whether or not the driver of the car should be acquitted, but the fact remains that everyone is capable of making this kind of mistake and of being guilty of nothing more than having the wrong expectations. A road that does not look like it is within the city limits gives rise to different expectations. You could even blame the local authorities for building a road that generates the wrong expectations among drivers. Sometimes I find myself wondering how fast I should be driving on a certain stretch of road. If there are no traffic signs to inform me, I usually try to guess the correct speed limit by looking at the features of the road on which I am traveling. What the road looks like determines the speed at which I think I am allowed to drive on it.

Given the very limited amount of information we are able to pick up from our surroundings, there is little point in increasing the number of traffic signs alongside the road. We also run the risk of overlooking important signs because we generally try to focus our attention primarily on the road. It would therefore be more efficient to meet drivers' expectations: if we see a road as a country road, we will be more inclined to drive faster on it than on a road that reminds us of a built-up area. We have previous experience with all kinds of roads, and we use that knowledge to reach our conclusions. We are more likely to interpret a road with several lanes as a highway, even when that road is in an urban area. Even the use of different kinds of traffic signs can lead to erroneous expectations: for example, we do not expect to see a sign within the city limits of New York telling us how many miles it is to Boston. A sign like that would only make people drive faster.

Our visual world is full of regularities, and we use this information when surveying our surroundings: knives are usually kept in the cutlery drawer, and that's why we are more inclined to look there rather than in the fridge when we need a knife. This is also the reason why a traffic sign on the left-hand side of the road (in countries that drive on the right) is less effective than a sign on the right. Context helps us to understand our complex visual world. So when we are driving, we have certain expectations and move our attention accordingly. A knife belongs in the cutlery drawer, and traffic signs should be on the right-hand side of the road. This is why it is unwise to display a billboard at a spot where drivers would normally expect to see a traffic sign. We have already seen in the example in chapter 5 how much time it costs to move your eyes to an object located off the road. Looking at a billboard on the side of the road can be just as dangerous, therefore, as glancing at your speedometer too often.

Our expectations as to where an object should be located in any given situation are based on previous experiences that can stretch back many years. But can these kinds of expectations also be created over a shorter period of time? To find out, scientists asked a group of test subjects to search for a target object, a *T*, for instance, among a large number of distractors. On each screen, the test subject was shown a number of objects arranged in a certain way: in the example in figure 6.1, the target object is located in the upper left-hand corner while the distractors are spread over the rest of the screen. What the test subjects did not know is that some of the screens were repeated and some were new.

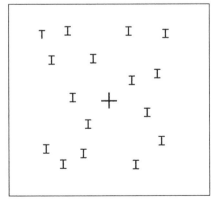

Figure 6.1
Search screen.

This experiment demonstrated that people are very sensitive to previous experiences when deciding where to direct their attention. The test subjects reacted more quickly to the target object when it appeared in a repeated screen than when it appeared in a new search screen. This effect is known as "contextual cueing": the context determines the spot to which attention will be guided. Measurements of eye movements in this experiment show that the eyes immediately move to the spot where the target object is expected to appear. Test subjects are usually unaware of this. After the experiment, most of them are unable to say which screens were repeated and which ones were new. So even though our attention is guided on the basis of this information, we do not have any conscious access to this knowledge. If you unconsciously leave your diary in the same spot on your desk often enough, you will never have a problem finding it regardless of how untidy your desk is.

The effects of context do not disappear immediately after the experiment either. When the test subjects in a contextual cueing experiment were asked to return one week later, they were still able to find the target object more quickly in the repeated screens. The test subjects were able to recall 60 individual screens, without being aware of the fact. It is also known that the effects are not limited to a few hours but can have an influence over much longer periods of time. The long-term effect of visual context was particularly evident in experiments in which test subjects were asked to search for a key in a scene from the real world, like a photo of a room or a mountain landscape. These experiments make use of the fact that we apparently have no difficulty recalling the details of hundreds of different scenes. The presence of a small key hidden in a scene still had a noticeable effect the following day. On the second day, the test subjects had to react as quickly as possible when they saw a target object (e.g., a small ball) in a scene; their reaction times were quickest when the target object was located in the same place as where they had found the key the day before. The most surprising discovery was probably the speed at which they moved their attention. The attentional shift was apparent even when the scene was shown very briefly. In previous chapters we talked about reflexive attentional shifts. In the case of the repeated scene, the attentional shift is so fast that it is almost like a reflexive attentional shift. In this kind of situation, and based on the content of the scene, we seem to know almost instantaneously that a certain location was important in a previous encounter because that was the spot where the key had been located. The following day, a reflexive attentional shift to that location occurs purely on the basis of this information. Our attention is guided by our experiences, and this often happens unconsciously.

Back to the traffic situation. Those charged with building new roads usually try to take the experiences of drivers into consideration in their designs. A well-designed road will not need any road signs at all and will be set up in such a way that drivers will understand how fast they should or should not drive on it. The less attention a driver has to pay to road signs, the more attention he or she will be able to pay to the road. The man credited with designing the modern roundabout, Hans Monderman, once said that the presence of road signs was a sure sign that the builder of the road had done a poor job. He also suggested that one way of getting drivers to slow down was to make the traffic situation less orderly, not more. In Italy, the police often park right in the middle of a busy intersection. This causes drivers to look twice and, consequently, to slow down. It seems counterintuitive to suggest that you can improve road safety by making the traffic situation more ambiguous, but we should remember that a road without obstacles only encourages drivers to step on the gas.

According to the principle of "shared space," it is better to design traffic spaces as "living spaces" by removing traffic-related objects like road signs, traffic lights, and curbs and replacing them with benches and flower boxes. This encourages road users to regulate the flow of traffic together with other users and stimulates a driver's sense of responsibility. These ideas have already been implemented in a town in the Netherlands, where they have led to a reduction in the number of traffic accidents.

When a particular stretch of road becomes the scene of regular traffic accidents caused by speeding, measures are often implemented that are aimed at encouraging drivers to automatically reduce their speed, such as narrowing the road or altering the road markings. Making the edges clearer is also a way of reinforcing the illusion of a narrow road. In built-up areas, the centerline that divides the road is sometimes removed to encourage drivers to be more aware of the need to share the road with oncoming traffic. When the centerline is left in place, drivers often forget that they need to be aware of oncoming traffic and increase their speed accordingly.

We have a terrific memory for the context in which objects are located. It appears that we are very good at remembering visual context because the information involved is of the unconscious kind, something for which we have an apparently unlimited memory. Everyone who can cycle a bike has the required skills stored away somewhere in his or her unconscious memory. Even though you cannot explain to someone else exactly how to ride a bike, you will have no problem riding one yourself even after a long winter out of the saddle. And this is also how it works with the unconscious information we gather about our visual world. We are able to store visual context without even knowing we have done so.

Repeating a certain visual context is of no benefit to people who have trouble picking up unconscious information, like learning a new motor skill. This includes patients with Parkinson's disease, who are unable to learn new unconscious motor skills as a result of problems with the basal ganglia. However, when the unconscious memory is still intact, as in the case of patients with Korsakoff's syndrome, experiments show that contextual cueing continues to function normally. Given that it is conscious memory that is affected in these patients, they will probably not be able to remember what they ate for breakfast but will still be able to react more quickly to a repeated search screen from the day before.

Notwithstanding their lack of conscious memory, the fact that Korsakoff patients still possess a well-functioning unconscious memory for visual context means that it can be used to learn new tasks. To do so, it is important that the information is acquired in a completely errorless manner. Otherwise the patients will also take the errors on board unconsciously, with the result that they will be unable to distinguish between a correct action and an erroneous one. Unfortunately, it is often assumed that patients who have no conscious memory are unable to learn new skills. As a result, many patients in nursing homes, for example, are denied this opportunity. Recent studies suggest that it is possible for them to acquire new skills when they use the "errorless learning" method. In a team of scientists led by Erik Oudman, we studied the errorless learning of a specific skill—how to operate a washing machine. This requires the ability to interact successfully with the external visual world by pushing the right buttons at the right time. Korsakoff patients who had never operated a washing machine before were able to do so after a few errorless learning sessions. They were unable to explain how they did it, however, because the required actions were not stored in their conscious memory. To them, it was like riding a bike. This shows that these patients still have learning potential when they make use of their almost unlimited unconscious memory. There is even evidence to suggest that this kind of memory has a much larger capacity and is far more robust than conscious memory—all the more reason to use it in our interaction with the visual world.

Our attention is influenced not only by the locations where we expect to find certain objects, but also by the visual associations we have with an object, like the way we associate "banana" with "yellow." In a study carried out by Chris Olivers, test subjects were asked to look for a particular road sign among a number of other road signs. All of the images of the signs were displayed in shades of gray, except for one distractor in full color that caused longer search times. The search time was even longer when the color of the distractor was associated with the identity of the target object, for example, red in the case of a stop sign. Even though color was not relevant to the task at hand, the test subjects' attention was guided by their association of "stop sign" with "red."

The strong effects of association may explain why some food manufacturers are prepared to take manufacturers of similar products to court for using the same color packaging as their own. In 2013, one case involved the similarities in color and design between two packages of rice. Well-known high-end brands invest a lot of money in associating their product with a certain color. Milk is generally associated with blue, even though the product itself is white. When we go looking for a carton of milk in the supermarket, we activate the color blue to speed up the process. A new player in the market, like a discount store, will use this association to ensure that the consumer's attention is drawn to their product. An orange carton of milk is unlikely to sell as well as a blue one. Court cases fought over similarity issues are rarely successful, given that the overall impression of the packaging of both products is also taken into account. Infringement of copyright will remain unproven in a case where the shape of the packaging or the shade of blue used on the competing product differs enough from the original. In the case of the packages of rice, much of the focus was on the resealable strip on the original product. The new product did not have a resealable strip, and so the judge decided that there was enough difference between the two products in terms of overall impression. When we go looking for a carton of milk, we don't have a specific shade of blue in mind; we're just looking for a carton that's blue.

Compare the packaging of the cheese spreads in figure 6.2. The two manufacturers produce the same product, and both use an image of a cheerful-looking cow with a meadow in the background. There are subtle differences in the shades of color used. Although it is not permitted to gain an advantage through association by imitating the packaging of a high-end product, there is also the matter of whether there is enough difference between the two products to avoid sowing confusion among consumers. In the example in figure 6.2, there are enough differences to ensure that the consumer does not confuse the two products, but as for their ability to attract attention, the effect will probably be the same for both.

Capturing attention on the basis of previous experience is known as "priming." You might be familiar with it from parties where you find yourself in the company of someone you really fancy. Your attention is focused fully on that person, and also on the clothes he or she is wearing. If the person goes home before you do, another person wearing the same color dress or shirt at the party will subsequently grab your attention automatically. And because your attention had been so focused on the red color of the person's dress or shirt, everything else in the room that happens to be red will attract your attention, too.

In experimental studies more attention has been paid recently to the influence the past has on the performance of test subjects. Visual attention experiments often involve the repetition of various kinds of tasks. One observation is never enough to provide the researcher with an accurate estimate of the reaction time for a particular task. Test subject "interference," like fluctuations in concentration or tiredness, means that multiple measurements are needed to arrive at a reliable estimate. This implies that

Figure 6.2
Association.

over the course of the experiment the past will play a role in how a task is performed. The reaction time for a particular task can also be heavily influenced by the reaction time for a previous task. For example, test subjects will find a target object much quicker when they are required to find the same target object as in a previous task.

Priming plays a particularly important role in relation to the aforementioned pop-out objects. When a unique color is repeated in a subsequent task, an object with the same color will be found more quickly. This also applies when a distractor suddenly has the same color as the unique object in the previous task. In that case, the distractor will have a significant effect on the performance of the test subject, whose attention will be captured by the distractor, meaning that the reaction time for the task will be relatively long. It is important to realize that not everyone reacts in the same way to a particular task; individuals are influenced by both their recent and distant past. This book will probably draw your attention in the bookstore more quickly if you have just seen a poster promoting it in the window.

How a person experiences the visual world is influenced by what he or she has seen and done in the past. Imagine you are lining up a series of domino tiles. After using only red tiles for a while, you will react a little slower when you are suddenly asked to continue with blue ones instead. Because you have been selecting only the red tiles, all of the other colors will have been temporarily suppressed. Your history of selecting only red tiles and ignoring all the other colors makes it more difficult to pick out a tile with a suppressed color. This is known as "negative priming."

Take a look at the illustration in figure 6.3. The test subject's task is to describe out loud and as quickly as possible the object in each image that has been drawn using an unbroken line. This is the dog in the image on the left and the trumpet in the image on the right. Performing this task correctly means selecting these objects at the expense of the object drawn using a dotted line. This is not difficult, but the consequences of our choice become apparent in the next task. When the target object is repeated (in this case the dog), priming will cause the reaction time to be shorter than it would be if the test subject had not seen the dog (positive priming). Conversely, the situation where the task involves ignoring the image of the dog results in negative priming: in this case the reaction time is longer than when the target object is new. This means that the representation of the visual image of the dog is lodged somewhere inside our brain. When the

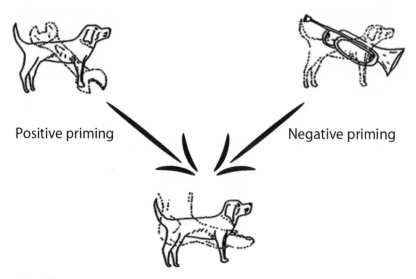

Positive priming   Negative priming

Figure 6.3
Priming.

representation is strong, we will identify the next dog quickly. But when the representation is suppressed, the reaction time will be significantly longer. The extra amount of time required to identify an object is an indication of the extent to which a representation is suppressed. Priming is very popular with illusionists: briefly exposing people to certain images increases the likelihood of their being influenced by those images when they are subsequently asked to draw something or to pick a card.

Memories influence our choice of where to move our attention. Our most vivid memories are probably those associated with events in which we were rewarded in some way or other. The spot on the street where I once found a 20-euro note will always remain vivid in my memory, and I often find my eyes being drawn automatically to that very spot whenever I am in the neighborhood. Okay, this incident is entirely anecdotal, but it is also possible to recreate this kind of situation in an experiment. Test subjects are told that they can earn money by performing a certain task well, but the size of the reward is determined by the presence of a particular color. When a green circle is visible during a search, the test subjects will earn more money when they find the target object than when the green circle is not present. In such a situation it is probably logical to assume that the color green will be very important to the test subjects and more than likely capture their attention. What is surprising, however, is that the effect remains the same even when the reward is no longer in play. Whereas the attention-grabbing effect during the reward phase can be explained as a kind of strategy on the part of test subjects (the color green is also relevant in terms of the amount of money that can be earned), this cannot explain the effect when the reward has been withdrawn. Nevertheless, when test subjects have associated a color with a reward for a certain period of time, that color will continue to automatically draw their attention, even in situations where the color acts as a distractor in a different search. Thanks to the test subjects' previous experience, the color has become a strong stimulus that cannot be easily suppressed, and the reaction times are longer than when the distractor is of a color that is associated with a lower reward. The effects were still apparent even weeks after the initial experiment.

Test subjects who are impulsive by nature were distracted for longer by the rewarding color. Less impulsive people find it easier to ignore objects that were initially associated with a reward. And although it may seem like a bit of a stretch to suggest this, it is possible that this has something to do with a person's susceptibility to addiction. When a certain drug has been responsible for causing a euphoric state of mind in the past, it will be more difficult to suppress the urge to relive that experience in a similar situation in the future. For example, a sign saying "Happy Hour" is bound to attract the attention of an alcoholic.

In the case of money, the reward is obvious enough. But even the prospect of new information can be seen as a reward, too. It appears that the brain's reward system becomes activated when, after completing a series of tasks with nothing but gray dots, we are presented with a task containing a red dot. When we experience something new, we see it as a kind of reward, just like money. Of course, we can think up all kinds of evolutionary reasons why this is so (e.g., our ancestors searching for food in new places), but it may also explain why we are so addicted to our mobile phones and computer screens today. They might just have some new information for us. The little envelope that appears in the bottom right-hand corner means that new information is available. Just try ignoring it. I, for one, find it almost impossible; if I don't turn off my e-mail, I will find myself checking it less than ten seconds after a new envelope has appeared on my screen. We are constantly on the lookout for new information. TV shows that provide a continuous stream of new information and multiple camera angles are more appealing to us than a program on repetitive content that uses only one camera angle, for example. We also love browsing through magazines because each new page offers us new information. An object that suddenly appears during a search assignment will automatically capture our attention because it has the potential to be relevant, but also simply because it offers new information.

Although phenomena like attention blindness and positive priming were only "discovered" by science within the last 40 years or so, magicians have been making use of our limited attentional capacity for centuries. It is often said that our experiments in the laboratory have very little to do with reality, but magicians have shown that there are plenty of attentional effects to be found in the real world, too. When performing their tricks, magicians constantly play with our expectations. A magician must be able to distract or move our attention to a spot where we think the action is playing out but where, in fact, nothing is happening. This allows the magician to produce new objects without the audience's noticing. If the audience knows where to look to see how a trick works, the trick won't work anymore.

Magicians use almost all of the tricks already alluded to in this book. They look away from the spot where a change is about to take place, click their fingers to distract our attention, and thwart our expectations by allowing changes to occur where we least expect them. The fact that we know they are making fools of us makes it all the more impressive and in no way diminishes the effectiveness of their tricks. A trick only ever fails to work when we know exactly what to look out for. In that case we focus our attention on the right spot, which allows us to see the change. It is a myth that magicians' tricks are all about speed and that objects disappear too fast for us to be able to notice. Of course, speed is important, but humans are unable to make something disappear so fast that other humans will not notice, provided they are paying attention. The trick lies in distracting our attention.

It is fascinating to see tricks that have been around for hundreds of years still being used in modern scientific experimental studies. One such experiment involved studying the eye movements of an audience watching a magician perform a trick in which he makes a cigarette "disappear" by letting it fall under a table while concentrating his gaze on and clicking his fingers. The results were very similar to the results of attention blindness experiments. The test subjects who failed to see the cigarette disappearing had seen the change with their own eyes but had not paid any attention to it.

Another good example is the trick with the disappearing ball. The magician throws a ball into the air a couple of times and it just seems to vanish suddenly midthrow. What happens is that on the final throw the magician makes it appear as if he has thrown the ball when in actual fact he still has it in his hand. He also follows the expected path of the ball with his head and eyes. To the audience, it looks like he has thrown the ball and that it just vanishes into thin air. In this case it is obvious that the audiences' eyes are looking at the right spot, but that their attention has moved to the expected location of the ball on the basis of the direction of the throw and where the magician is looking.

Whenever a magician produces a dove and lets it fly off, you can be sure that an important change has taken place somewhere else. Magicians know that when there is movement in several places at the same time, the audience will focus its attention on the most salient movement. All of a magician's actions, body language, eye movements, and speech have a function. A well-timed joke can distract attention from a trick, which makes it less likely that the change will be noticed. Memory also plays an important role in the success of a magic trick. Magicians know that they should never perform the same trick twice for the same audience. The chances that the audience will spot the change the second time around are much greater, even if they don't know where that change will take place. Audiences make a connection between an action and a change, which makes them less prone to distraction and attention blindness.

Today, training your sharpness of perception is big business. The organizations that offer this service draw their inspiration from studies that show that it is possible to improve our perception by repeating a simple task. The effects of "perceptual learning" have been proven to exist in experimental settings, but there is still some doubt as to the influence they have on our daily lives. This is because the effects of perceptual learning don't usually manifest themselves in other situations. In studies of perceptual learning, test subjects spend several hours a day, and several days in succession, training themselves to perform a simple visual task: distinguishing a certain color or orientation, for example. If you train yourself long enough, you will inevitably get better at distinguishing a particular color or orientation.

This in itself is a little surprising because you would expect that our years of experience with the visual world would have made experts out of us already. It appears that training can make you even better than you already were. However, training does not make you better at distinguishing *other* colors or orientations. The effect applies to one specific skill: for example, you may become an expert in detecting vertical motion, but that will not make you any better at detecting the horizontal kind. This expertise in a particular motion is known to last, and the effects of perceptual learning are often still noticeable months later. Perceptual learning is an unconscious process, however, meaning that test subjects are unable to say exactly what they have learned.

One example of a study that has become popular with organizations in the perceptual learning business involves an experiment in which a group of baseball players received training in the perception of objects. The training consisted of thirty 25-minute sessions in which the test subjects had to detect a barely visible object. After the training, their perception had improved and they even scored more points during subsequent baseball games. Their ball-strike rate improved, and they were also better at deciding when not to swing the bat. Excellent results at first glance, but the study also throws up a few problems. For instance, there was no control group, which makes it unclear as to whether the effects could just as easily be attributed to a placebo effect. Of course, you could also say that this is how it always works with science: someone proposes an idea, and other scientists attempt to study it further. However, it becomes problematic when organizations start to make far-reaching conclusions on the basis of a handful of studies. They promise great things: thanks to the development of portable screens and apps, it is becoming easier to do a short training session each day that will improve your perceptive powers. But the scientific evidence for the effects of this kind of visual training is very thin on the ground.

Another recent innovation on the market is stroboscopic glasses: professional eyewear that is aimed at helping soccer players, for example, to react faster to a ball. The technique is already proving very popular with clubs all around the world. The idea behind the glasses is quite simple: instead of projecting the visual image without interruption, it is blocked 5 to 150 times per second. The glasses produce a flickering image so that instead of unbroken movement the user sees a series of stationary images. The inventor of stroboscopic eyewear claims that it improves the speed of action of the wearer because of our ability to process stationary objects more quickly than moving ones. A flickering image should, in theory, improve a goal-keeper's ability to spot an approaching ball. The effects of training with the glasses may also help in situations where the players are not wearing them, as the brain is then trained to glue stationary images together to achieve a flowing movement. As of yet, there is little scientific evidence to support these claims. Although users are enthusiastic and report significant improvements, the lack of reliable control groups means that it is impossible to make any scientific claims regarding the effectiveness of these kinds of training methods.

If tightly controlled experiments show that the effects of perceptual learning do not necessarily apply to other situations, the question is whether stroboscopic eyewear training is of any real benefit. There has been a remarkable upsurge in the number of visual training methods available on the market, methods whose mechanism may appeal to the imagination but that have very little scientific basis. For the moment, all we can do is wait until the required scientific studies into these techniques have been carried out.

The older we get, the poorer our perception becomes. This is primarily the result of changes in the eye, such as loss of flexibility in the lens, and is not related to changes in the brain. There are no visual training methods that can improve the condition of the eye itself, but there are ways in which you can improve your interaction with your visual environment. One such method is the training aimed at increasing the size of the attentional field that we discussed in chapter 3. This method focuses on visual attention, as opposed to perception, and this may be the crucial reason why attention training can be beneficial.

Of course, none of this means that there is no such thing as perceptual learning. For example, as children we learn to distinguish our *V*s from our *U*s. Much of the visual expertise that we develop is based on perceptual learning. I still find it amazing how quickly I am able to interpret experimental data, while as a student I often found myself staring helplessly at charts in the vain hope of understanding what they were trying to tell me. The ability to quickly review data has been made more difficult by the emergence of "big data" in which it is hard to identify patterns simply because of the overabundance of data. We have to rely on computer algorithms to find the patterns for us and are unable to identify patterns ourselves by taking "a quick look" at the data, like chemists do when they want to decipher a chemical formula. Ask an expert why he or she sees a particular pattern, and the expert won't be able to give you an answer; it's all done automatically. The pioneer of perceptual learning, Eleanor J. Gibson, once wrote: "We don't just see, we look; we don't just hear, we listen." Pilots are able to check the status of their airplane very quickly by looking at the plane's gauges. They don't inspect each individual gauge, however, because that would require their attention and take far too long. It's all about being able to distinguish visual patterns without shifting your attention—relying on your "gut feeling" in other words.

This is an example of how perceptual learning works in practice. It also applies in the medical world where surgeons learn how to identify patterns by interpreting images of body tissue. When a gall bladder is removed, for instance, a camera is used to provide images from inside the body. The surgeon uses the camera to navigate his or her way to the right spot. It takes a long time to develop the required expertise, but perceptual learning can help to accelerate that process. In an experiment, one half of a group of students were shown a series of images very briefly and then asked to identify the parts of the body the images came from. The other students were shown the images for a longer period of time and were allowed to navigate for as long as they liked. In the exams, the perceptual learning group scored four times better than the group that had been given more time to look at the images. This demonstrates that developing an instinct for identifying patterns can be extremely effective. Techniques aimed at enabling automatic identification of patterns are now being used in medical training to help students to quickly identify various types of skin conditions, for example.

Our memory determines how we experience the world around us. You could say that we look at the present through the spectacles of the past. This is quite logical, because we have learned where to look to find the information that interests us, and it helps us to react quickly to certain situations and to select the right information from our surroundings for further processing. When our expectations are not fulfilled, however, we are confronted with the limits of our perception and can easily miss a scooter speeding toward us in the middle of town.

7 What Happens When It All Goes Wrong? What Brain Damage Can Teach Us about Attention

Seventy-four-year-old Hank has been a painter all his life. Even after retiring he still busies himself every day with his paint and brushes. One day, Hank is painting his brother's fence when he falls off the ladder after suffering a massive stroke. In the hospital the neuropsychologist asks him if he knows why he is in the hospital. Hank thinks for a moment and says: "Because an ambulance brought me here."

"Are you in any pain?"

"Not really," he replies. "I'm just a bit tired and the food here is terrible."

What he doesn't tell the doctor is that he is ignoring half of his visual world. Because of the damage done to his parietal lobe, Hank is now suffering from "visual neglect," a condition whereby patients experience problems moving attention to the left or the right side of the visual world.

Neglect usually results from damage to the right hemisphere of the brain. The attention regions in that part of the brain are responsible for moving attention to the left visual field. The problems that arise from neglect can be extremely disabling for the patient, but at the same time fascinating because they show us just how important attention is to exploring the world around us. The condition has different levels of severity, and patients with the most severe form are completely unaware of what goes on in the neglected half of their world. When Hank has his dinner, he only eats the food on the right-hand side of his plate. When he is finished eating, he believes that he has eaten everything because he has no access to the information on the other side of the plate. Only when his plate is turned around does the other half of his meal appear in the "intact" part of his visual world and does he realize that he hasn't finished his food after all. When he brushes his teeth, he only does so in the part of his visual field that is still intact. And as for shaving, he only ever does the right-hand side of his face.

The most difficult aspect of this condition is that the majority of patients have no idea they are suffering from it. In his conversation with the neuropsychologist, Hank says that the reason he is in the hospital is because an ambulance brought him there; he says nothing about any problems with moving his attention. He is missing half of his world, but he just hasn't noticed. This lack of insight into your own condition makes recovery very difficult. The aim of the conversation with the neuropsychologist is to broaden that insight. She attempts to move Hank's attention to the damaged side. So she asks him more questions. "Look at your left hand. What do you see?"

Thanks to the neuropsychologist's clear instructions, Hank is able to move his attention to the left. He sees his left hand. He looks at the hand and says: "This hand is clean. It usually has paint on it." He then switches his gaze quickly to his right hand. "This one hurts because of the injection the doctor gave me." His attention has returned to his right hand, and his left hand promptly ceases to exist again. This is a typical scenario with neglect patients: they are actually able to move their attention, but only after receiving clear instructions and for a short period of time. Their attention shifts swiftly back to that part of the visual field that is not being neglected. If the door to the room is on his left-hand side, a doctor can walk in without Hank's noticing. Only when the doctor calls his name will he slowly start to move his attention toward the doctor.

We have already seen that attention plays a crucial role in establishing the identity of an object in the world. If patients pay no attention to a certain part of the world, they will not notice any of the objects located there. This may explain the lack of insight into their own condition: how would you react if someone told you that you were missing a whole section of the visual world and that you didn't even know it? You probably wouldn't believe them. That is why the role of a neuropsychologist is so important. By making Hank aware of the objects he is missing, he will hopefully realize that he is neglecting a part of his world.

Around 25 percent of all patients with brain damage suffer from some form of neglect. Fortunately, it is usually a short-term problem. This is because there are all kinds of processes in the brain that are disrupted in the acute phase, but that are eventually able to return to normal. After a stroke, for example, excess blood has to be drained off from the brain. When that is done, many brain functions return to normal and the problem of neglect just vanishes. Even within only a few days of suffering brain damage, a patient may show no more signs of neglect. However, for some patients neglect remains a chronic condition, meaning that the problems they have with moving their attention are permanent.

There are a number of short tests that can be done to find out if patients are suffering from neglect, tests that patients can do from their hospital bed. One example is the Line Bisection Test in which the patient is asked to divide a number of lines on a sheet of paper precisely in two. To be able to pinpoint the middle of the line, the patient needs to be aware of the whole line. Someone suffering from left-side neglect will have great difficulty with this test and will probably divide the line on the right-hand side. By ignoring the left side of the visual world, the patient obviously ignores the left-hand side of the line, too, which makes the line appear shorter and, consequently, the middle of the line is judged to be more to the right.

Look at the example in figure 7.1. The top line is marked precisely in the middle. The bottom line shows how a patient with left-side neglect carries out this task; the mark has clearly shifted to the right.

Simple copying exercises are also used to identify the disorder. For example, a patient suffering from neglect will only draw half a flower or half a clock. Drawing a clock, in particular, throws up some interesting examples, like the one in figure 7.2. Most neglect patients will draw a full circle because drawing a circle is an automatic motion, and it is almost impossible not to finish drawing one once you've started. It does not require any attention. Patients also add all of the numbers to the clock, but not where you would expect them to be drawn. They know that there are twelve numbers on a clock, but they have trouble trying to squeeze them all in.

Figure 7.1
Middle of the line.

Figure 7.2
Neglect.

The act of searching also presents problems. If a patient with left-side neglect is asked to cross out all the stars on a sheet of paper containing lots of distractors, he or she will only cross out the ones on the right-hand side. Most of the stars on the left will be ignored. When the patient's eye movements are measured, the neglect becomes very obvious: he or she makes almost no eye movements to the left. This is not because of some motor skill problem; the lack of eye movements to the left is because of a lack of attention for that side. If a patient has no attention for the left-hand side, there is simply no reason for the patient to make an eye movement to the left visual field.

There are many different forms of neglect, including neglect with specific consequences for near or far visual space, neglect for auditory or visual information, neglect for a part of the body, and neglect for the left or right side of objects. An interesting form of neglect found in some patients is the neglect of parts of images recalled from memory. For example, when asked to describe the town square in their place of birth from a certain perspective, some patients with left-side neglect will only describe the right-hand side of the square. However, when they are asked to stand on the other side of the square in their mind, they will describe the opposite side (the right-hand side from this new angle). This shows that neglect can have an effect not only on visual images as perceived by the eyes, but also on imagined images. There are even cases in which patients exhibit neglect only for imagined images and not for visual ones.

It is interesting to note that after a certain period of time many patients develop strategies to compensate for their attentional deficit. As a result, their performance during standard neuropsychological tests, like the Line Bisection Test and drawing tests, displays no aberrations when compared to the performance of people who do not suffer from neglect. However, this does not mean that they no longer suffer from an attentional disorder. It is possible that a patient in the chronic phase will still show symptoms of neglect when he or she is required to perform a task within a limited period of time or under pressure. At such moments, the patient will not be able to use the compensation strategies that eradicated the mistakes during standard tests. Tests on patients in the chronic phase who are required to perform difficult tasks often reveal a mild form of neglect. The question, therefore, is just how well a patient in the chronic phase can be said to have recovered. Walking down a busy street is a much more complicated task for the brain than marking the middle of a line from the relative comfort of a hospital bed.

It is important not to confuse neglect with blindness. Patients with neglect can see the entire visual world, but they have difficulty moving their attention. It is different for people with cortical blindness. Cortical blindness can arise from brain damage, but it usually affects the visual cortex, the region where basic visual information is processed. Damage to the right hemisphere can also cause problems in the left visual field, and vice versa. The blindness almost never affects the whole visual field, but just the part that is processed by the neural regions that have suffered damage.

Patients with cortical blindness are aware of their problem very soon after suffering damage. They make eye movements to their blind side more frequently because they know that they may be missing information on that side. They are also usually able to draw a complete flower and cross out all the stars in a neuropsychological test. These patients often use a white cane when out walking, but they are only blind to a certain part of the visual field. Cortical blindness has nothing to do with the eyes, but rather with those parts of the brain that are responsible for processing visual information. Patients with left-sided blindness often tell me that they are having problems with their left eye, only to realize, after closing that eye, that the blindness is also affecting the left-hand side of their right eye. The blindness is in a part of the visual field that is perceived by both eyes. Unlike neglect, cortical blindness is not an attentional deficit. There is no basic visual information in the blind field to which patients can move their attention. Unlike patients with neglect, people who suffer from cortical blindness cannot see any colors, shapes, or other visual building blocks in the affected field.

You can probably imagine how difficult it is sometimes for a doctor to distinguish between neglect and cortical blindness. In both cases, patients are missing information in the affected visual field, either through a lack of attention or a lack of visual processing. Picture the following situation: a patient is asked to focus his eyes on a dot. At the same time, a red square appears briefly in the left visual veld. The patient's task is to describe what he has detected. Both a patient with neglect and a patient with cortical blindness will probably say that they did not detect anything. The only way to distinguish between the two disorders in a situation in which no eye movements are made is to keep the red square in view. A patient with cortical blindness will still be unable to detect the object, regardless of how long it remains there. A patient with neglect, on the other hand, will be able to shift his attention to the affected field, albeit very slowly. It is much easier to distinguish between the two based on eye movements: the patient with cortical blindness will make an eye movement to the left, because he knows that there may be information located there. This does not happen in the case of neglect in the acute phase.

There is a lot of work being done on developing rehabilitation methods for these disorders. In the case of cortical blindness, it is often assumed that the disorder is a permanent one with no prospect of recovery. There are a number of software packages on the market that claim to be able to reduce the size of the blind area, but these are not in general use in rehabilitation centers. Over the past few years, a number of studies have suggested that in some situations a training program could help to alleviate this blindness. In the training, the patient has to distinguish multiple objects in the blind visual field. The improvements observed in some patients are concentrated primarily on the edges of the blind area. Sometimes the gains are very small, but in some cases the blindness is reduced by half.

In the case of neglect, there are many different training methods, but there have not been enough controlled studies with sufficient numbers of patients to test their effectiveness. An additional problem is that there are many different types of neglect, and that the various methods probably only work for a particular type. Determining which method helps which type of neglect takes a lot of time, not to mention financial support, usually from a large pharmaceutical company, to fund the research. However, initial results are promising. For example, prism glasses can help to reduce neglect in some patients. The lenses in these glasses are polished in such a way that they can bend the visual world so that the incoming visual information appears at a different spot to where the information is actually located. After wearing these glasses for a while, patients are able to pick up objects from a table without grasping at thin air. At that point they are said to be "adapted." When they remove the glasses, however, the adaptation wears off very quickly, but long-term effects have also been achieved through multiple sessions. It appears that the prism adaptation is able to reset the attention system. This may sound a little vague, but at the moment it is the best explanation available. Additional studies will have to be carried out to see which patients can benefit from this method and how prism adaptation works exactly.

Some patients with cortical blindness are able to react to information in the blind field. In their case, the blindness appears to be relative. For example, there are patients who are able to correctly state above chance level which light out of a set of four has been illuminated in their blind field. This is not an easy task for a patient. Imagine if I were to ask you over the phone how many fingers I am holding up. You would probably refuse to answer the question. How could you be expected to know? However, when these patients are asked to guess which light has been switched on, they generally score above the 25 percent chance level. The same results were achieved in relation to movement, whereby patients were able to correctly state whether a particular movement was downward or upward, while at the same time indicating that they were unable to see the actual movement. These phenomena have also been reported for other building blocks related to perception, such as color and orientation. It is interesting to note that these patients never actually say that they have detected the color green or red, for example. This information is only revealed when they have to guess the color. When an object in the blind visual field does have an effect on the patient's behavior, this is called "blindsight."

So how does this work? To be able to see, we process visual information in the visual cortex in the rear of our brain. Damage to this region can result in cortical blindness. However, the brain is a vast network, and so the visual information can be found in other parts of the brain, too. Some of these regions receive the visual information but do not process it to the same level as the visual cortex. Unlike what happens in the visual cortex, the processing that goes on in these regions does not lead to conscious perception. For example, the information gets sent to the superior colliculus, a region in the midbrain that is largely responsible for determining where we focus our eyes. Experiments have shown that a salient distractor in the blind visual field of a patient with cortical blindness can have an effect on eye movements. What we don't know, however, is whether this is actually of any benefit to the patient or just another interesting discovery made in a lab. It is possible that a salient object that appears suddenly in a patient's blind field could result in a reflexive eye movement toward that object more quickly than in the case of a patient for whom an object in the blind field has no effect.

In the case of blindsight, the knowledge relating to the incoming visual information is available in the brain, but it cannot be accessed in a conscious manner by the patient. It is worth emphasizing that this applies to special cases only: these effects are not seen in the majority of patients with cortical blindness (although it could also be the case that the right experiments have yet to be carried out). It is not clear why this affects some patients and not others. Studies have shown that the strength of the connections between the eyes and the midbrain may play a role, but it has also been suggested that the damage to the visual cortex in these patients is limited and that there may still be "islands" of functioning neurons, meaning that reflex actions based on the incoming information are still possible, despite the absence of conscious perception.

The research into blindsight raises an interesting question: what is the influence of information that people cannot say they have detected but that has definitely been registered by the eyes? We already know that we are only able to report a limited amount of the visual information that falls on the retina. Let's refer to this visual information as "consciousness." You are "conscious" of only a very limited part of all the visual information you receive. The rest of that information is processed in the brain, but you remain unaware of it, just like in the case of blindsight. An interesting phenomenon that can help to explain the influence of this unconscious information is the so-called attentional blink. The attentional blink occurs in situations where a stream of information needs to be processed very quickly. For example, a test subject may be asked to report two red letters in a stream of black letters. The letters are displayed one after the other at the same spot in the middle of the screen, but then for only 100 milliseconds.

As this is long enough to allow test subjects to distinguish individual letters, they usually report the first red letter correctly. What happens with the second red letter is more interesting, however. When it appears directly after the first letter, the test subject usually has no trouble reporting it, but when a few black letters (up to about five black letters) appear between the first and second red letters, most of the test subjects are unable to report the second letter. And when there are more than five black letters between the two red letters, the results are normal again.

This means that our attention is shut down briefly when we have to store a letter in our working memory. Not only must test subjects identify the first red letter, they also have to be able to recall it at the end of the information stream because they will be asked to report which two red letters they detected. In the time it takes to store the first red letter in their memory, no other letters can be identified; it's as if the attention has blinked briefly (hence the term "attentional blink"). It takes a moment for the attention to start blinking, however, and this explains why it is possible to store a red letter that appears directly after the first red letter. The interesting thing about these results is that the second red letter can escape the attention of the test subject even though it is presented for as long as 100 milliseconds on an otherwise empty screen. Of course, it would be more accurate to say that it escapes the conscious perception of the test subject. After all, the second letter has actually been processed by a part of the brain, but the test subject is still unable to report the second red letter.

To what extent is the second object processed during this rapid stream of information? Up to now we have been dealing with a red letter appearing between black letters, but when you change the identity of the second target object to such a degree that it becomes important information for test subjects, their attention will blink a lot less. It appears that we do have access to the identity of the second target object, but only in situations in which the information in the blink is important to us. This might all sound rather dualistic, but it appears that the decision as to when normally unconscious information is allowed to permeate our consciousness is made *for* us and not by us. When, for example, test subjects are required to search for a certain color within a stream of information, the emotional overtone of the second word influences the size of the attentional blink. Words that are regarded as negative, such as "war," are reported as a second target object more often than words without any negative overtones. All of the words on the screen are processed, therefore, at the semantic level so that the test subject's brain knows the meaning of a word. This knowledge is required to be able to determine whether the word is important or not. A similar reduction in the size of the attentional blink can be found in a variation on this task in which the search involves faces with particular expressions. Faces with an angry expression during the attentional blink were reported more often than faces with an emotionally neutral expression. The same applies to images of poisonous spiders, with the reduction being even greater for people with a fear of spiders.

It may be useful to take a step back for a moment and address the relationship between attention and consciousness again. This is a controversial topic and one that inspires regular scientific debate. In my view, the results with respect to the attentional blink mean that information that cannot be reported is processed nonetheless, even to a level that allows the brain to establish the meaning of the information. The meaning of the object in the attentional blink subsequently determines whether the object is registered at the conscious or unconscious level. We have already seen that attention is required to be able to establish the identity of an object. This means that attention is a crucial precondition for consciousness, and we can only be conscious of attended information. However, we are not conscious of all the information that has been attended. After our attention has been moved to a particular object and its identity has been established, the decision still has to be made whether this information becomes conscious or unconscious.

Let's look at the example of the attentional blink again. The entire stream of words is presented in the middle of the screen. So there is no doubt as to where the attention is focused. All of the words that are presented one after another will fall within the attentional window. This also explains why all of the words are processed up to the semantic level. However, during the blink we are not conscious of the words that are being shown on the screen. The term "attentional blink" may, in fact, be slightly misleading: it is not attention that blinks, but your consciousness.

There are different ways of reducing the attentional blink. Together with her colleagues, the Dutch scientist Heleen Slagter discovered that meditation can lead to a reduced attentional blink. Test subjects took part in a three-month-long, intensive training in *vipassana*, a form of meditation that that uses many of the techniques associated with mindfulness. Compared with a group of test subjects who had never had any training in meditation, they became much better at identifying the second target object after the meditation training program. Although the test subjects were not actually meditating during the task, they did perform better because of their previous experience with meditation. This experience meant that they were able to remain conscious of all the information despite the need to store the identity of the first target object in their memory.

Meditation teaches you how to react less rigidly to incoming visual information, which in turn makes you less likely to ignore other information. In experiments where test subjects did not have to focus all their attention on finding the two target objects, the attentional blink proved to be reduced. In fact, if you encourage test subjects to perform extremely well by offering them some kind of reward, their performance will actually deteriorate. In that case they will concentrate too hard on the task and close the gate too quickly after seeing the first target object, which often causes them to miss the second object. The best approach is just to relax and not to worry too much about your performance. You are more open to the outside world when you are in a relaxed frame of mind. This information is worth remembering; you never know when you might be able to put it into practice.

The effect of concentration on the strength of the attentional blink probably explains the large differences sometimes found within a group of test subjects with regard to attentional blink. It is also known that not every test subject has an attentional blink. When performing this task, so-called nonblinkers exhibit brain activity different from that of test subjects who do have an attentional blink. They appear to be quicker at processing information and therefore more flexible about allowing information to become conscious. Other studies have even linked nonblinking with character traits like openness and extroversion. On the other hand, people with a tendency to act neurotically are more likely to have a stronger attentional blink. And although we might be entering a field here in which terms are less well-defined than in the research into visual perception, these studies demonstrate that the consciousness filter does not work the same way for everyone.

It is not easy to apply the results of the research into the attentional blink to the real world. And yet the real world is a place where we are continually bombarded with visual information that the brain must choose to make conscious or not. Studies have shown that the relevance of the information is crucial: we become conscious of relevant information, and we ignore the rest.

If it is possible to process unconscious information up to the semantic level, then it must also be possible to influence people at the unconscious level. To discuss this further, we must first review the concept of visual masking. Visual masking experiments involve the brief presentation of information, like a word. The test subject is not aware of the word because new visual information, such as a number of x's at the spot where the letters were presented (a so-called mask), is displayed immediately after the word has disappeared. This mask blots out the briefly shown word so that the test subject is not able to become conscious of it. When the mask disappears, a new, clearly visible word is displayed, and the test subject's task is to indicate whether this word is spelled correctly or not. When the meaning of the masked word is related to the subsequent word, the test subject reacts much faster than when the words are unrelated (e.g., when the word "doctor" follows the masked word "nurse," the reaction is much faster than when the masked word is "bread"). This effect is known as subliminal (or unconscious) priming.

Subliminal priming is the dream of every advertising agent. Who wouldn't want to be able to influence consumers without their even noticing? You can get consumers to do almost anything simply by presenting information at random moments in such a way that they do not notice it. Not surprisingly, this kind of advertising is prohibited in many countries.

But does subliminal priming actually work? You may have heard of the infamous Coca-Cola case from 1957 in which it was alleged that a marketing man, James Vicary, had inserted the words "drink coke" and "eat popcorn" into a movie that was showing in a movie theater in New Jersey. It was alleged that the method used prevented the moviegoers from consciously perceiving the messages. The words appeared on-screen for only 1/30th of a second. According to Vicary, sales of the products shot up as a result: Coca-Cola by 18 percent and popcorn by a whopping 58 percent. It was front-page news all over America, and the whole country was up in arms. Later on, it turned out that the story was a complete fabrication. Vicary had made it all up in an effort to draw attention to his marketing firm. Attempts were made later to execute the same kind of study in a conventional manner, but most of the tests were unable to prove whether subliminal messages had any effect on the behavior of viewers. There is one exception, however: information that is acquired subliminally does appear to have an effect when the message is in harmony with the test subject's objective. For example, a phrase like "drink coke" will only work with test subjects who are actually thirsty and whose objective it is to find something to drink (in this experiment the test subjects had been given something salty to eat beforehand). However, it is unlikely that messages like these have any real effect in advertisements. It is worth pointing out that these experiments were carried out in a very short space of time. The effects are probably temporary and may disappear altogether a few minutes after the attempt to influence the viewer has been made.

Nevertheless, there are plenty of other conspiratorial tales regarding subliminal messaging. For example, in a political broadcast by George W. Bush during the Bush versus Gore election campaign, the word "rats" was shown very briefly on its own after the name Al Gore, as part of the line "bureaucrats decide." The word appeared a little earlier than the rest of the line but the makers of the clip denied all wrongdoing and said that it had not been done on purpose. However, this did not prevent all hell from breaking loose. In 2007 there was a similar case in the United States involving the TV show *Iron Chef* when the logo of McDonalds (one of the show's sponsors) appeared very briefly in an episode of the program. Once again, the makers denied everything and blamed the incident on a technical malfunction.

Some movie directors openly admit to the use of subliminal messaging and even see it as a kind of personal trademark. The movie *Fight Club*, for example, is full of subliminal images that do not have any obvious purpose. The makers of the movie apparently had a lot of trouble convincing the movie company's quality controllers that the images had been deliberately inserted into the movie and were not the result of human error.

In subliminal priming the information is displayed very briefly before being masked. There are also techniques that allow the information to be displayed for a few seconds without the test subject's noticing. These techniques make use of the fact that both of our eyes usually look at the same image. By using mirror images and two screens, one can get each eye to look at a different image. At first, the viewer will see only one image without any overlap. After a while, however, the second image will slowly become visible, presenting the viewer with two alternating images. This leads to a rivalry between the images, which then have to compete with each other (the phenomenon is known as "binocular rivalry"). You could, of course, make sure that one of the images is so strong that it will always win out over the other one. An example of this is when you have a moving image (e.g., black-and-white blobs) in one eye and a static image in the other. Even when you look at these images for a few seconds, you will never be able to report the content of the static image. This is interesting because one of the images is in full view for a few seconds and is projected to one eye, which makes it similar to the situation faced by patients with cortical blindness, but then minus the damage to the brain.

This technique makes it possible to study the effect of subliminal information presented to test subjects who are free of brain damage. One of the most intriguing examples concerns an experiment in which erotic images were presented to the "suppressed" eye. The test subjects were subsequently unable to report what they had seen, apart from the moving blobs. However, they still reacted more quickly to a target object when it was presented in the same spot as where the erotic image was presented. This faster reaction was evident in men only when the suppressed erotic image had been that of a naked woman, and worked for women only when it had been a naked man. The extent to which the images were processed was determined, therefore, by the sexual preference of the test subject. Of course, not all men reacted faster after seeing a naked woman; gay men only reacted faster after they had seen a picture of a naked man.

By making a small adjustment to this technique, the suppressed image can eventually become visible to the test subject. This involves making the suppressed image gradually stronger. At first the image is weak, but after a second it is completely clear. The suppressed image is then able to break through the image of the blobs. This technique can be used to study how long it takes for an image to permeate visual consciousness. Important information always gets through faster than unimportant information. And once again, it is our brain that decides which information will be given priority on the basis of the content of the image.

For example, we have discovered that it is the content of our working memory that determines how much priority is given to an image. We asked a number of test subjects to remember a color. With this color safely stored in their memory, we then presented an image to their suppressed eye. When the color of this image was the same as the color the test subjects had stored in their working memory, the permeation time was a lot shorter than when the color was different. It appears that our brain is constantly busy prioritizing relevant information. Think back to the example of the party where you couldn't get the person dressed in red out of your mind. In that case, not only is your attention captured by all the red clothing you see, but every other scrap of red information will permeate your visual consciousness much quicker, too.

The behavior of our painter, Hank, will also be influenced by the information in the neglected field. A good example of the influence of information in the neglected field comes from an experiment in which two line drawings of a house, one under the other, were shown to a patient with left-side neglect. She had to say which house she would prefer to live in. The houses were identical, except for one detail: flames were coming out of the house on the bottom but not out of the one on top. The flames were drawn on the left-hand side of the house, so the patient was not aware of them. However, when asked which house she would like to live in, the patient replied that she would prefer to live in the top one (the one that was not on fire). Although she did not know why, her choice was clear without her having to actually say that it wouldn't be nice to live in a house that was on fire.

With regard to the influence of neglected information, it is known that for neglect the situation is quite similar to what happens with the attentional blink: in some patients, the information is processed up to the semantic level. For example, presenting the image of an apple very briefly in the neglected field elicits a faster reaction to the word "tree" when it is shown in the intact field, compared to an image that has nothing to do with the tree at the semantic level. This means that the building blocks of visual information in the neglected field are, in fact, bound and that the patient's brain does have access to the identity of the neglected object. Neglect should not be regarded, therefore, as the impossibility of moving attention to the neglected field. When no competing information is presented in the intact field, the sudden appearance of information in the neglected field can result in an attentional shift, but the information will not permeate consciousness. In this case there is enough attention to gain access to the identity of an object, but not enough to allow it to permeate the patient's consciousness. It is worth pointing out that this only works in situations where no other visual information is being presented. An image in the intact visual field that is presented at the same time as an image in the neglected field will inevitably draw all of the attention to the intact field and prevent any further processing from taking place in the neglected field.

So That's How Attention Works

Hank's situation clearly illustrates what happens to information to which we pay no attention. Because of the limits on our ability to move our attention, we are not able to register the details of every single moment in our external visual world. This may explain why attention architects are prepared to fight so hard for our attention. After all, whoever has our attention also has the possibility of gaining access to our consciousness, even if only for a fraction of a second. The rest of the information is simply ignored, and there is nothing worse for an attention architect than for people to ignore his or her message.

I still think that the spotlight in a theater is one of the best metaphors you can use to explain attention. The comparison rings true, except for one crucial detail: in the theater we don't get to operate the spotlight ourselves, but with attention we are the ones at the controls. In some cases, however, we are the victims of our own reflexes, and our attention becomes automatically drawn to intrusive information from the outside world. Sometimes this cannot be helped, but even just knowing that our attention can be captured in this way is very useful. For instance, when you really need to concentrate on something, you can make your spotlight small enough to be able to ignore intrusive information from your surroundings. There is a limit to the extent to which you can ignore everything around you, however. Your reflexes are tasked with reacting to dangerous situations, for example. Unfortunately, these reflexes can also be "misused" to allow intrusive information to get through to us. All around us screens compete for our attention, and it is those attention architects who know the most about how our attention works who tend to come out on top.

There is a lot of talk these days about the effect that the increasingly busy visual world is having on our ability to concentrate. It is claimed, for example, that children are not able to focus their attention on anything for very long anymore because there is just too much incoming information. This has not yet been proven scientifically, however, and the truth may not be as clear-cut as we think. It may be that we are actually getting better at guiding our attention because we are becoming more and more used to ignoring intrusive information. Take banner blindness, for instance, and how good we have become at dealing with flashing banners on websites. We all know that it is not easy to find your way around at first in a new environment where there are lots of distractions, but when you are in familiar surroundings, previous experiences help you to know where you are, despite all the distractions around you. So it appears to be far too simplistic to make any definitive statement about our failing attention: it depends entirely upon the situation in which you find yourself. Experiences and expectations also differ depending on the situation. In any event, we are certainly not slaves to our reflexive attention. We are able to take control, and we know how to use rapid attentional shifts to navigate our way efficiently through the world and ignore intrusive information.

You don't need to be a trend watcher to be able to predict that in the future our eye movements will be monitored even more closely. The emergence of eye trackers will give attention architects access to information on what we like to look at. This represents a veritable gold mine of information. Whoever knows what we look at also knows what interests us. You won't even have to ask a person this anymore; all you need to do is track his or her eye movements. Of course, being able to relay information to someone requires more than just enticing the person to look your way, but it is a crucial precondition. No one is ever going to buy your product if people are not looking at your advertisement or if the product is displayed in the wrong place in the store.

This is how attention works. It filters incoming information. But how does this affect the way we experience our visual world? After you put down this book, have a good look around you. What information is available? If our access to the visual world is so limited, to what extent is our picture of that world, the movie theater in which we sit, an accurate representation of the world around us? Difficult to say, but in any event it is accurate enough to create the impression that the world around us is actually our world. We know when to move our attention to the right place at the right time. The light in the fridge is always on. But it is when things go wrong that we realize that our ability to see the entire picture is just an illusion. Remember the barrier at the tunnel at the start of this book? How on earth could you miss it?

Acknowledgments

In 1890, the founding father of contemporary experimental psychology, William James, wrote, "Everyone knows what attention is." Although this may be true in terms of how we use the word "attention" in everyday language ("My attention keeps wandering..."), it does not apply to the scientific study of attention. This book represents a humble attempt to explain what current scientific research tells us about *visual attention*. However, chances are that much of that knowledge will be completely outdated a hundred years from now. And that's exactly how it should be. Science is a dynamic process in which new discoveries are built on established knowledge. It is even possible that some of the findings in this book have already passed their sell-by date. After all, they are highly colored by my own vision, and not every claim I make will go unquestioned. My main objective, however, was to write a book that is accessible to everyone. This means, of course, that I couldn't address every single nuance or contradiction. Science is never finished. But that does not mean that it is a futile pursuit. On the contrary, each new piece of knowledge that we acquire along the way brings us closer to discovering how things work and helps us to better understand the world around us.

While there may be broad consensus over much of what is written in this book, that does not apply to everything in it. When I decided to write a book that would appeal to an audience beyond the scientific community, it was not my aim to paint a detailed picture of the current scientific debate, but rather to explain what our knowledge of attention can tell us about our daily lives. I know that some of my colleagues will frown upon some of what I have written, but I hope that I have managed to drum up some enthusiasm among readers for the questions that we are continually trying to answer and for the beauty of the experiments we carry out. After all, that is what I love most about my profession: devising experiments in order to better understand our behavior. Studies of patients with brain damage augment this work and help us to understand what goes wrong when the system becomes damaged.

You may have noticed that much of the research referred to in this book was conducted in the Netherlands. This is no coincidence, of course, as a lot of it was carried out by my own colleagues and is therefore very familiar to me. This is not the only reason for my use of Dutch sources, however. The Netherlands is a world leader in experimental psychology and neuropsychology. We have a strong tradition in the field and a vanguard of dedicated scientists who keep the flame burning brightly.

A number of my colleagues were kind enough to run their eye over large sections of the book. In this regard I would particularly like to thank my PhD supervisor, Jan Theeuwes. He encouraged me to be critical of my own findings and tried to steer me away from making pronouncements that were in any way doubtful, scientifically speaking. I hope, for his sake, that I have found the right balance in this respect. I also borrowed many of the practical examples in the book from him. Jan Theeuwes is the head of the cognitive psychology department at the VU University in Amsterdam, a place renowned for its groundbreaking research into visual attention. I will be forever grateful for having been able to complete my PhD there. Thanks also to the other attention expert at the VU, Chris Olivers, for all his inspiration and for checking the various chapters.

My colleagues at the Helmholtz Institute in Utrecht were also kind enough to help with proofreading the book. In particular, I would like to thank Ignace Hooge and Chris Paffen. They were quick to help when I was in danger of losing the plot in relation to basic perception. Ignace also provided me with many excellent examples of the everyday implications of our research. The Helmholtz Institute is a terrific place to work. Thank you to Albert Postma for supporting me. And thanks to Tanja Nijboer for all the fun we have doing our research together.

Thanks also to Frans Verstraten and Victor Lamme for their positive criticism of the first draft of the book.

I would like to thank everyone at the AttentionLab for the superb research that we do, the great ideas that we have yet to explore, the pure enjoyment I get from working together, and the many lessons that I learn from you every day. The same goes for all the students I have had the pleasure of tutoring over the years.

I am very grateful to Eva van den Broek for recommending me to a fantastic publisher while my book was still in its "vague idea" phase. The people at Maven have been extremely helpful in publishing the original Dutch version of this book. Sander Ruys saw something in my idea, and together we came up with a plan. Emma Punt supervised the editing and pointed out the weak parts where I was likely to bore the reader. I just find it hard sometimes to believe that others might not be as interested as I am in reading about yet another experiment into yet another phenomenon. I now have a better understanding of this, fortunately. Many thanks to Lydia Busstra for the promotional work on this book.

For the international edition of this book, I would like to thank Danny Guinan for the great work on the translation and Anne-Marie Bono and Katherine Almeida at MIT Press. It's an honor to have my book published by MIT Press.

I have many wonderful international collaborators, friends and tutors who have sharpened my research, helped me understand visual attention and make this work so much fun: Bob Rafal, Jason Barton, Mike Dodd, Andrew Hollingworth, Doug Munoz, David Melcher, Wieske van Zoest, Clayton Hickey, Tod Braver, and Masud Husain. I thank Janet Bultitude for her friendship and the inspiration for Hank, the painter in chapter 7.

Jannie brought me cups of tea every night up in the attic, which also gave her a chance to actually speak to me every now and then. After 17 years together she knows what I'm like. Fortunately, I also had Jasper and Merel to drag me away from my scribbling every now and then. They can never demand too much attention from me (except on Sunday mornings, maybe). I am very proud of both of them.

I would also like to take the opportunity to thank my parents and my sister for all they have done for me. My gratitude knows no bounds.

In the end it is you, the reader, who will decide whether this book has achieved what I hoped it would. So don't hesitate to let me know what you think. In the meantime, thank you kindly for your attention.

Notes

Epigraphy

James, William. (1890). *Principles of psychology* (p. 403). New York, NY: Holt.

Preface

Inattentional Blindness while Pursuing a Suspect
Chabris, C., Fontaine, M., Simmons, D., & Weinberger, A. (2011). You do not talk about fight club if you do not notice fight club: Inattentional blindness for a simulated real-world assault. *i-Perception, 2,* 150–153.

Chapter 1

The World as an External Hard Drive
O'Regan, J. K. (1992). Solving the "real" mysteries of visual perception: The world as an outside memory. *Canadian Journal of Psychology, 46,* 461–488.

The Rapid Processing of Scenes
Potter, M. C. (1976). Short-term conceptual memory for pictures. *Journal of Experimental Psychology: Learning, Memory, and Cognition, 2*(5), 509–522.

Ideas on Infobesity Borrowed from a Lecture Given by Prof. Chris Olivers
Olivers, C. (2014). *iJunkie: Attractions and distractions in human information processing.* VU University, Amsterdam.

How Virtual Reality Can Adapt to Our Perception
Reingold, E. M., Loschky, L. C., McConkie, G. W., & Stampe, D. M. (2003). Gaze-contingent multiresolutional displays: An integrative review. *Human Factors, 45*(2), 307–328.

Chapter 2

Train Crash at Ladbroke Grove
UK Health and Safety Commission. (2000). The Ladbroke Grove Rail Inquiry.

I am grateful to Ignace Hooge for everything he taught me about visibility and salience. He also kindly gave me permission to use many examples from his own lectures to students of psychology at Utrecht University.

The Visibility and Salience of Fire Trucks
Solomon, S. S., & King, J. G. (1997). Fire truck visibility. *Ergonomics in Design*, 5(2), 4–10.

The Success of the Third Stop Lamp
Kahane, C. J. (1998). The long-term effectiveness of center high mounted stop lamps in passenger cars and light trucks. *NHTSA Technical Report Number DOT HS*, 808, 696.

"The Dress" and Color Constancy
Lafer-Sousa, R., Hermann, K. L., & Conway, B. R. (2015). Striking individual differences in color perception uncovered by "the dress" photograph. *Current Biology*, 25, R1–R2.

The Clearview Typeface
Garvey, P. M., Pietrucha, M. T., & Meeker, D. (1997). Effects of font and capitalization on legibility of guide signs. *Transportation Research Record*, 1605, 73–79.

Macular Degeneration
Van der Stigchel, S., Bethlehem, R. A. I., Klein, B. P., Berendschot, T. T. J. M., Nijboer, T. C. W., & Dumoulin, S. O. (2013). Macular degeneration affects eye movement behaviour during visual search. *Frontiers in Perception Science*, 4, 579.

Reading Ability of Patients with Macular Degeneration
Falkenberg, H. K., Rubin, G. S., & Bex, P. J. (2006). Acuity, crowding, reading and fixation stability. *Vision Research*, 47(1), 126–135.

Eccentric Viewing Spectacles
Verezen, Anton. Eccentric viewing spectacles. Thesis, Radboud University, Nijmegen.

Compulsory Eye Tests
Jan Theeuwes had a convincing article about compulsory eye tests published in the Dutch newspaper *NRC Handelsblad* on July 9, 2004 in which he claimed that opticians would be the only ones to benefit from such eye tests.

Wolfe, J. M., Kluender, K. R., Levi, D. M., Bartoshuk, L. M., Herz, R. S., Klatzky, R., et al. (2015). *Sensation & Perception*. Sunderland, MA: Sinauer Associates.

The "Sensation & Perception" course in the psychology degree course at Utrecht University makes use of this book as a general reference for topics like the blind spot, the physiology of the retina and the regions of the brain responsible for processing the building blocks

Chapter 3

The Sensitivity of X-rays in Public Health Screenings
Setz-Pels, Wikke. Improving screening mammography in the south of the Netherlands: Using an extended data collection on diagnostic procedures and outcome parameters. Thesis, Erasmus University, Rotterdam.

Radiologists and the Invisible Gorilla
Drew, T., Vo, M. L.-H., & Wolfe, J. M. (2013). The invisible gorilla strikes again: Sustained inattentional blindness in expert observers. *Psychological Science*, 24(9), 1848–1853.

The Invisible Guide Wire in Multiple Scans
Lum, T. E., Fairbanks, R. J., Pennington, E. C., & Zwemer, F. L. (2005). Profiles in patient safety: Misplaced femoral line guidewire and multiple failures to detect the foreign body on chest radiography. *Academic Emergency Medicine*, 12(7), 658–662.

The Difference between Drillers and Scanners
Drew, T., Vo, M. L.-H., Olwal, A., Jacobsen, F., Seltzer, S. E., & Wolfe, J. M. (2013). Scanners and drillers: Characterizing expert visual search through volumetric images. *Journal of Vision*, 13(10), 3.

Search Methods of Security Scan Operators
Biggs, A. T., Cain, M. S., Clark, K., Darling, E. F., & Mitroff, S. R. (2013). Assessing visual search performance differences between Transportation Security Administration Officers and nonprofessional visual searchers. *Visual Cognition*, 21(3), 330–352.

Wolfe, J. M., Brunelli, D. N., Rubinstein, J., & Horowitz, T. S. (2013). Prevalence effects in newly trained airport checkpoint screeners: Trained observers miss rare targets, too. *Journal of Vision*, 13(3), 33.

Mitroff, S. R., Biggs, A. T., Adamo, S. H., Wu Dowd, E., Winkle, J., & Clark, K. (2014). What can 1 billion trials tell us about visual search? *Journal of Experimental Psychology: Human Perception and Performance*, 41(1), 1–5.

Attention as the Binder of Building Blocks
Treisman, A. M., & Gelade, G. (1980). A feature-integration theory of attention. *Cognitive Psychology*, 12, 97–136.

Incorrectly Bound Visual Objects

Treisman, A., & Schmidt, H. (1982). Illusory conjunctions in the perception of objects. *Cognitive Psychology*, 14, 107–141.

Friedman-Hill, S. R., Robertson, L. C., & Treisman, A. (1995). Parietal contributions to visual feature binding: Evidence from a patient with bilateral lesions. *Science*, 269, 853–855.

Useful Field of View Training

Ball, K., Berch, D. B., Helmers, K. F., Jobe, J. B., & Leveck, M. D. (2002). Effects of cognitive training interventions with older adults: A randomized controlled trial. *Journal of the American Medical Association*, 288(18), 2271–2281.

Inattentional Blindness

Mack, A., & Rock, I. (1998). *Inattentional blindness*. Cambridge, MA: MIT Press.

Inattentional Blindness in Pilots with Head-Up Displays

Haines, R. F. (1989). A breakdown in simultaneous information processing. In G. Obrecht & L. W. Stark (Eds.), *Presbyopia research: From molecular biology to visual adaptation* (pp. 171–175). New York, NY: Plenum Press.

Change Blindness

O'Regan, J. K., Rensink, R. A., & Clark, J. J. (1999). Change-blindness as a result of "mudsplashes." *Nature*, 398(6722), 34.

Change Blindness in Eyewitness Accounts

Nelson, K. J., Laney, C., Bowman Fowler, N., Knowles, E. D., Davis, D., & Loftus, E. F. (2011). Change blindness can cause mistaken eyewitness identification. *Legal and Criminological Psychology*, 16, 62–74.

Change Blindness in Movies

Smith, T. J., & Henderson, J. M. (2008). Edit blindness: The relationship between attention and global change blindness in dynamic scenes. *Journal of Eye Movement Research*, 2(2), 6.

Chapter 4

Pop-outs

Itti, L., & Koch, C. (2001). Computational modelling of visual attention. *Nature Reviews. Neuroscience*, 2(3), 194–203.

Automatic Capture of Attention

Theeuwes, J. (1992). Perceptual selectivity for color and form. *Perception & Psychophysics*, 51, 599–606.

Automatic Capture of Attention in a Cockpit
Nikolic, M. I., Orr, J. M., & Sarter, N. B. (2004). Why pilots miss the green box: How display context undermines attention capture. *International Journal of Aviation Psychology*, 14(1), 39–52.

Automatic Capture of Attention by Faces
Langton, S. R. H., Law, A. S., Burton, A. M., & Schweinberger, S. R. (2008). Attention capture by faces. *Cognition*, 107(1), 330–342.

Automatic Capture of Attention by Spiders
Lipp, O. V., & Waters, A. M. (2007). When danger lurks in the background: Attentional capture by animal fear-relevant distractors is specific and selectively enhanced by animal fear. *Emotion*, 7(1), 192–200.

Automatic Capture of Attention by a Painful Object
Schmidt, L. J., Belopolsky, A., & Theeuwes, J. (2015). Attentional capture by signals of threat. *Cognition and Emotion*, 29(4), 687–694.

Mulckhuyse, M., Crombez, G., & Van der Stigchel, S. (2013). Conditioned fear modulates visual selection. *Emotion*, 13(3), 529–536.

Disengaging from Your Own Face
Devue, C., Van der Stigchel, S., Brédart, S., & Theeuwes, J. (2009). You do not find your own face faster; you just look at it longer. *Cognition*, 111(1), 114–122.

Selective Searching for Objects with Specific Features
Egeth, H., Virzi, R. A., & Garbart, H. (1984). Searching for conjunctively defined targets. *Journal of Experimental Psychology: Human Perception and Performance*, 10, 32–39.

Automatic and Voluntary Shifting of Attention
Posner, M. I., & Cohen, Y. (1984). Components of visual orienting. In H. Bouma & D. G. Bouwhuis (Eds.), *Attention and performance X: Control of language processes*, 531–556. Hillsdale, NJ: Lawrence Erlbaum Associates.

Shifting Attention through a Facial Cue
How do we know whether facial cues really lead to an automatic shift? The best experimental manipulation used to demonstrate this was also a very elegant one. In the experiment the facial cue was not 50 percent valid with regard to the location of the target object, but only 25 percent. This means that in most cases the face provided an opposing indication and the test subjects would have good reason not to shift their attention in the direction indicated by the face. And yet, the test subjects could not resist the urge to do the opposite. If a target object is presented immediately after a misleading face is shown, the test subject's attention will still shift to the spot where the face is looking. Only when there is a longer gap between the appearance of the face and the appearance of the target object is it possible to shift one's attention more quickly to the spot where the face is not looking.

Driver, J., Davis, G., Ricciardelli, P., Kidd, P., Maxwell, E., & Baron-Cohen, S. (1999). Gaze perception triggers reflexive visuospatial orienting. *Visual Cognition*, 6(5), 509–540.

Friesen, C. K., Ristic, J., & Kingstone, A. (2004). Attentional effects of counterpredictive gaze and arrow cues. *Journal of Experimental Psychology: Human Perception and Performance*, 30, 319–329.

Shifting Attention as a Result of a Facial Cue among Children with Autism

Senju, A., Tojo, Y., Dairoku, H., & Hasegawa, T. (2004). Reflexive orienting in response to eye gaze and an arrow in children with and without autism. *Journal of Child Psychology and Psychiatry, and Allied Disciplines*, 45(3), 445–458.

Frischen, A., Bayliss, A. P., & Tipper, S. P. (2007). Gaze cueing of attention: Visual attention, social cognition, and individual differences. *Psychological Bulletin*, 133(4), 694–724.

The Influence of Emotion on a Face in Facial Cues

Terburg, D., Aarts, H., Putman, P., & Van Honk, J. (2012). In the eye of the beholder: Reduced threat-bias and increased gaze-imitation towards reward in relation to trait anger. *PLoS One*, 7(2), e31373.

The Influence of Political Preference on the Effect of Facial Cues

Dodd, M. D., Hibbing, J. R., & Smith, K. B. (2011). The politics of attention: Gaze-cueing effects are moderated by political temperament. *Attention, Perception & Psychophysics*, 73(1), 24–29.

Shifting Attention by Numbers

Dodd, M. D., Van der Stigchel, S., Leghari, M. A., Fung, G., & Kingstone, A. (2008). Attentional SNARC: There's something special about numbers (let us count the ways). *Cognition*, 108(3), 810–818.

Shifting Attention by Arrows

Hommel, B., Pratt, J., Colzato, L., & Godijn, R. (2001). Symbolic control of visual attention. *Psychological Science*, 12(5), 360–365.

Shifting Attention by Acquired Cues

Dodd, M. D., & Wilson, D. (2009). Training attention: Interactions between central cues and reflexive attention. *Visual Cognition*, 17(5), 736–754.

Chapter 5

The Visual Stability of Our World despite Our Eye Movements

Cavanagh, P., Hunt, A. R., Afraz, A., & Rolfs, M. (2010). Visual stability based on remapping of attention pointers. *Trends in Cognitive Sciences*, 14(4), 147–153.

Burr, D., & Morrone, M. C. (2011). Spatiotopic coding and remapping in humans. *Philosophical Transactions of the Royal Society of London. Series B, Biological Sciences,* 366, 504–515.

The Regions of the Brain Responsible for Updating Relevant Objects during Eye Movements
Duhamel, J.-R., Colby, C. L., & Goldberg, M. E. (1992). The updating of the representation of visual space in parietal cortex by intended eye movements. *Science,* 255, 90–92.

Walker, M. F., Fitzgibbon, E. J., & Goldberg, M. E. (1995). Neurons in the monkey superior colliculus predict the visual result of impending saccadic eye movements. *Journal of Neurophysiology,* 73(5), 1988–2003.

The Antisaccade Task
Everling, S., & Fischer, B. (1998). The antisaccade: A review of basic research and clinical studies. *Neuropsychologia,* 36(9), 885–899.

The Effect of Damage to the Frontal Areas on the Antisaccade Task
Pierrot-Deseilligny, C., Muri, R. M., Ploner, C. J., Gaymard, B. M., Demeret, S., & Rivaud-Pechoux, S. (2003). Decisional role of the dorsolateral prefrontal cortex in ocular motor behaviour. *Brain,* 126(6), 1460–1473.

The Antisaccade Task and ADHD
Munoz, D. P., Armstrong, I. T., Hampton, K. A., & Moore, K. D. (2003). Altered control of visual fixation and saccadic eye movements in attention-deficit hyperactivity disorder. *Journal of Neurophysiology,* 90, 503–514.

Rommelse, N. N. J., Van der Stigchel, S., & Sergeant, J. A. (2008). A review on eye movement studies in childhood and adolescent psychiatry. *Brain and Cognition,* 68(3), 391–414.

The Effect of Training on the Antisaccade Task
Dyckman, K. A., & McDowell, J. E. (2005). Behavioral plasticity of antisaccade performance following daily practice. *Experimental Brain Research,* 162(1), 63–69.

The Behavior of Professional Athletes during the Antisaccade Task
Lenoir, M., Crevits, L., Goethals, M., Wildenbeest, J., & Musch, E. (2000). Are better eye movements an advantage in ball games? A study of prosaccadic and antisaccadic eye movements. *Perceptual and Motor Skills,* 91, 546–552.

The Influence of a Positive Frame of Mind on the Antisaccade Task
Van der Stigchel, S., Imants, P., & Ridderinkhof, K. R. (2011). Positive affect increases cognitive control in the antisaccade task. *Brain and Cognition,* 75(2), 177–181.

The Antisaccade Task and Schizophrenia
Sereno, A. B., & Holzman, P. S. (1995). Antisaccades and smooth pursuit eye movements in schizophrenia. *Biological Psychiatry*, 37, 394–401.

The Relationship between Attention and Eye Movements
Rizzolatti, G., Riggio, L., Dascola, I., & Umilta, C. (1987). Reorienting attention across the horizontal and vertical meridians: Evidence in favor of a premotor theory of attention. *Neuropsychologia*, 25, 31–40.

Van der Stigchel, S., & Theeuwes, J. (2007). The relationship between covert and overt attention in endogenous cueing. *Perception & Psychophysics*, 69(5), 719–731.

Banner Blindness
Benway, J. P. (1998). Banner blindness: The irony of attention grabbing on the World Wide Web. *Proceedings of the Human Factors and Ergonomics Society Annual Meeting*, 42(5), 463–467.

The Usefulness of Studying the Eye Movements of Experts
Lichfield, D., Ball, L. J., Donovan, T., Manning, D. J., & Crawford, T. (2010). Viewing another person's eye movements improves identification of pulmonary nodules in chest x-ray inspection. *Journal of Experimental Psychology. Applied*, 16, 251–262.

Mackenzie, A., & Harris, J. (2015). Using experts' eye movements to influence scanning behaviour in novice drivers. *Journal of Vision*, 15, 367.

Subtle Gaze Manipulation
Sridharan, S., Bailey, R., McNamara, A., & Grimm, C. (2012). Subtle gaze manipulation for improved mammography training. In *Proceedings of the ACM SIGGRAPH Symposium on Applied Perception in Graphics and Visualization*, 75–82.

The Classification of Clinical Groups on the Basis of Eye Movements
Tseng, P.-H., Cameron, I. G. M., Pari, G., Reynolds, J. N., Munoz, D. P., & Itti, L. (2012). High-throughput classification of clinical populations from natural viewing eye movements. *Journal of Neurology*, 260, 275–284.

Chapter 6

Failing to Spot a Police Car on the Hard Shoulder
Langham, M., Hole, G., Edwards, J., & O'Neill, C. (2002). An analysis of "looked but failed to see" accidents involving parked police vehicles. *Ergonomics*, 45, 167–185.

Failing to Spot Cyclists on the Road
Theeuwes, J., & Hagenzieker, M. P. (1993). Visual search of traffic scenes: On the effect of location expectations. In A. Gale (Ed.), *Vision in vehicle IV*, 149–158. Amsterdam: Elsevier.

Contextual Cueing during Visual Searches
Chun, M. M. (2000). Contextual cueing of visual attention. *Trends in Cognitive Sciences*, 4(5), 170–178.

Eye Movements and Contextual Cueing
Peterson, M. S., & Kramer, A. F. (2001). Attentional guidance of the eyes by contextual information and abrupt onsets. *Perception & Psychophysics*, 63(7), 1239–1249.

The Capacity of Contextual Cueing
Jiang, Y., Song, J.-H., & Rigas, A. (2005). High-capacity spatial contextual memory. *Psychonomic Bulletin & Review*, 12(3), 524–529.

The Influence of Long-Term Memory on Our Attention
Summerfield, J. J., Lepsien, J., Gitelman, D. R., Mesulam, M. M., & Nobre, A. C. (2006). Orienting attention based on long-term memory experience. *Neuron*, 49, 905–916.

Contextual Cueing in Patients with Parkinson's Disease and Korsakoff's Syndrome
Van Asselen, M., Almeida, I., Andre, R., Januario, C., Freire Goncalves, A., & Castelo-Branco, M. (2009). The role of the basal ganglia in implicit contextual learning: A study of Parkinson's disease. *Neuropsychologia*, 47, 1269–1273.

Oudman, E., Van der Stigchel, S., Wester, A. J., Kessels, R. P. C., & Postma, A. (2011). Intact memory for implicit contextual information in Korsakoff's amnesia. *Neuropsychologia*, 49, 2848–2855.

How Patients with Korsakoff's Syndrome Can Learn How to Operate a Washing Machine
Oudman, E., Nijboer, T. C. W., Postma, A., Wijnia, J., Kerklaan, S., Lindsen, K., et al. (2013). Acquisition of an instrumental activity of daily living in patients with Korsakoff's syndrome: A comparison of trial and error and errorless learning. *Neuropsychological Rehabilitation*, 23(6), 888–913.

The Capacity and Robustness of Unconscious Memory
Lewicki, P., Hill, T., & Bizot, E. (1988). Acquisition of procedural knowledge about a pattern of stimuli that cannot be articulated. *Cognitive Psychology*, 20, 24–37.

Reber, A. S. (1989). Implicit learning and tacit knowledge. *Journal of Experimental Psychology. General*, 118, 219–235.

The Influence of Associations on Visual Searching
Olivers, C. N. L. (2011). Long-term visual associations affect attentional guidance. *Acta Psychologica*, 137, 243–247.

The Effects of Priming on Reaction Times

Kristjánsson, Á., & Campana, G. (2010). Where perception meets memory: A review of repetition priming in visual search tasks. *Attention, Perception & Psychophysics, 72,* 5–18.

Meeter, M., & Van der Stigchel, S. (2013). Visual priming through a boost of the target signal: Evidence from saccadic landing positions. *Attention, Perception & Psychophysics, 75,* 1336–1341.

Negative Priming

Mayr, S., & Buchner, A. (2007). Negative priming as a memory phenomenon—A review of 20 years of negative priming research. *Zeitschrift für Psychologie/Journal of Psychology, 215,* 35–51.

The Automatic Influence of Reward on Visual Attention

Anderson, B. A., Laurent, P. A., & Yantis, S. (2011). Value-driven attentional capture. *Proceedings of the National Academy of Sciences of the United States of America, 108*(25), 10367–10371.

Ideas on the Reward for New Information Borrowed from a Lecture Given by Prof. Chris Olivers

Olivers, C. (2014). iJunkie: Attractions and distractions in human information processing. VU University, Amsterdam.

The Influence of New Information on Our Reward System

Biederman, I., & Vessel, E. A. (2006). Perceptual pleasure and the brain. *American Scientist, 94,* 249–255.

The Science of Magicians

Macknik, S. L., King, M., Randi, J., Robbins, A., Teller, Thompson, J., & Martinez-Conde, S. (2008). Attention and awareness in stage magic: Turning tricks into research. *Nature Reviews Neuroscience, 9,* 871–879.

Kuhn, G., & Land, M. F. (2006). There's more to magic than meets the eye. *Current Biology, 16*(22), R950–R951.

Perceptual Learning

Gibson, E. J. (1988a). Exploratory behavior in the development of perceiving, acting, and the acquiring of knowledge. *Annual Review of Psychology, 39,* 1–41.

Fahle, M. (2005). Perceptual learning: Specificity versus generalization. *Current Opinion in Neurobiology, 15*(2), 154–160.

The Influence of Perceptual Learning on Baseball Players

Deveau, J., Ozer, D. J., & Seitz, A. R. (2014). Improved vision and on-field performance in baseball through perceptual learning. *Current Biology, 24*(4), R146–R147.

Sebastiaan Mathôt discusses this study in his blog: http://www.cogsci.nl/blog/miscellaneous/226-can-you-brain-train-your-way-to-perfect-eyesight/. The article on stroboscopic glasses appeared on August 29, 2015, in *de Volkskrant*.

Perceptual Learning in the Medical World

Guerlain, S., Brook Green, K., LaFollette, M., Mersch, T. C., Mitchell, B. A., Reed Poole, G., et al. (2004). Improving surgical pattern recognition through repetitive viewing of video clips. *IEEE Transactions on Systems, Man, and Cybernetics*, 34(6), 699–707.

Chapter 7

An Excellent Source for More Information on Neglect

Driver, J., & Mattingley, J. B. (1998). Parietal neglect and visual awareness. *Nature Neuroscience*, 1(1), 17–22.

How Patients with Neglect Draw a Clock

Di Pellegrino, G. (1995). Clock-drawing in a case of left visuospatial neglect: A deficit of disengagement. *Neuropsychologia*, 33(3), 353–358.

Eye Movements in Patients with Neglect

Behrmann, M., Watt, S., Black, S. E., & Barton, J. J. S. (1997). Impaired visual search in patients with unilateral neglect: An oculographic analysis. *Neuropsychologia*, 35(11), 1445–1458.

Van der Stigchel, S., & Nijboer, T. C. W. (2010). The imbalance of oculomotor capture in unilateral visual neglect. *Consciousness and Cognition*, 19(1), 186–197.

Neglect in Far and Near Space

Van der Stoep, N., Visser-Meily, A., Kappelle, L. J., De Kort, P. L. M., Huisman, K. D., Eijsackers, A. L. H., et al. (2013). Exploring near and far regions of space: Distance-specific visuospatial neglect after stroke. *Journal of Clinical and Experimental Neuropsychology*, 35(8), 799–811.

Neglect of Imagined Images

Cuariglia, C., Padavani, A., Pantano, P., & Pizzamiglio, L. (1993). Unilateral neglect restricted to visual imagery. *Nature*, 364, 235–237.

Nys, G. M., Nijboer, T. C. W., & De Haan, E. H. (2008). Incomplete ipsilesional hallucinations in a patient with neglect. *Cortex*, 44, 350–352.

Subtle Neglect in Patients in the Chronic Phase

Bonato, M., & Deouell, L. Y. (2013). Hemispatial neglect: Computer-based testing allows more sensitive quantification of attentional disorders and recovery and might lead to better evaluation of rehabilitation. *Frontiers in Human Neuroscience*, 7, 162.

The Difference between Neglect and Cortical Blindness
Walker, R., Findlay, J. M., Young, A. W., & Welch, J. (1991). Disentangling neglect and hemianopia. *Neuropsychologia*, 29(10), 1019–1027.

The Effect of Training on Cortical Blindness
Bergsma, D. P., & Van der Wildt, G. J. (2010). Visual training of cerebral blindness patients gradually enlarges the visual field. *British Journal of Ophthalmology*, 94, 88–96.

Sabel, B. A., & Kasten, E. (2000). Restoration of vision by training of residual functions. *Current Opinion in Ophthalmology*, 11, 430–436.

The Effects of Prism Adaptation in Patients with Neglect
Nijboer, T. C. W., Nys, G. M. S., Van der Smagt, M., Van der Stigchel, S., & Dijkerman, H. C. (2011). Repetitive long-term prism adaptation permanently improves the detection of contralesional visual stimuli in a patient with chronic neglect. *Cortex*, 47(6), 734–740.

Nijboer, T. C. W., Olthoff, L., Van der Stigchel, S., & Visser-Meily, A. (2014). Prism adaptation improves postural imbalance in neglect patients. *Neuroreport*, 25(5), 307–311.

The Influence of "Blind" Visual Information on Patients with Cortical Blindness
Van der Stigchel, S., Van Zoest, W., Theeuwes, J., & Barton, J. J. S. (2008). The influence of "blind" distractors on eye movement trajectories in visual hemifield defects. *Journal of Cognitive Neuroscience*, 20(11), 2025–2036.

Ten Brink, T., Nijboer, T. C. W., Bergsma, D. P., Barton, J. J. S., & Van der Stigchel, S. (2015). Lack of multisensory integration in hemianopia: No influence of visual stimuli on aurally guided saccades to the blind hemifield. *PLoS One*, 10(4), e0122054.

The Different Kinds of Blindsight
Weiskrantz, L. (1986). *Blindsight: A case study and implications*. Oxford: Oxford University Press.

Attentional Blink
Raymond, J. E., Shapiro, K. L., & Arnell, K. M. (1992). Temporary suppression of visual processing in an RSVP task: An attentional blink? *Journal of Experimental Psychology: Human Perception and Performance*, 18, 849–860.

Olivers, C. N. L., Van der Stigchel, S., & Hulleman, J. (2007). Spreading the sparing: Against a limited-capacity account of the attentional blink. *Psychological Research*, 71(2), 126–139.

The Influence of Emotional Words on the Attentional Blink
Anderson, A. K., & Phelps, E. A. (2001). Lesions of the human amygdala impair enhanced perception of emotionally salient events. *Nature*, 411, 305–309.

The Influence of Meditation on the Attentional BlinkS
lagter, H. A., Lutz, A., Greischar, L. L., Francis, A. D., Nieuwenhuis, S., Davis, J. M., et al. (2007). Mental training affects distribution of limited brain resources. *PLoS Biology*, 5(6), e138.

The Influence of Irrelevant Thoughts and Music on the Attentional Blink
Olivers, C. N. L., & Nieuwenhuis, S. (2005). The beneficial effect of concurrent task-irrelevant mental activity on temporal attention. *Psychological Science*, 16(4), 265–269.

Individual Differences in the Attentional Blink
Martens, S., Munneke, J., Smid, H., & Johnson, A. (2006). Quick minds don't blink: Electrophysiological correlates of individual differences in attentional selection. *Journal of Cognitive Neuroscience*, 18(9), 1423–1438.

MacLean, M. H., & Arnell, K. M. (2010). Personality predicts temporal attention costs in the attentional blink paradigm. *Psychonomic Bulletin & Review*, 17(4), 556–562.

The Role of Unconscious Information on Our Choice of Soda
Karremans, J. C., Stroebe, W., & Claus, J. (2006). Beyond Vicary's fantasies: The impact of subliminal priming and brand choice. *Journal of Experimental Social Psychology*, 42(6), 792–798.

The Role of Sexual Preference in Suppressed Images
Jiang, Y., Costello, P., Fang, F., Huang, M., & He, S. (2006). A gender- and sexual orientation-dependent spatial attentional effect of invisible images. *Proceedings of the National Academy of Sciences of the United States of America*, 103, 17048–17052.

The Interaction between Working Memory and Visual Consciousness
Gayet, S., Paffen, C. L. E., & Van der Stigchel, S. (2013). Information matching the content of visual working memory is prioritized for conscious access. *Psychological Science*, 24(12), 2472–2480.

Gayet, S., Van der Stigchel, S., & Paffen, C. L. E. (2014). Breaking continuous flash suppression: Competing for consciousness on the pre-semantic battlefield. *Frontiers in Psychology*, 5(460), 1–10.

The Effect of Drawings of a House in Flames on Patients with Neglect
Marshall, J. C., & Halligan, P. W. (1988). Blindsight and insight in visuo-spatial neglect. *Nature*, 336(2), 766–767.

The Processing of Information in the Neglected Field
McGlinchey-Berroth, R., Milberg, W. P., Verfaellie, M., Alexander, M., & Kilduff, P. (1993). Semantic priming in the neglected field: Evidence from a lexical decision task. *Cognitive Neuropsychology*, 10, 79–108.

Index